Anatomy Review

CONTENTS

To

Gayle, Albert and Paul

ANATOMY REVIEW

Fourth Edition

EDITED BY

MARVIN I. GOTTLIEB, Ph.D., M.D.
Former Instructor of Anatomy
The University of Tennessee
Medical Units Division
Supervisor of Clinical Services
Child Development Center

1500 MULTIPLE CHOICE QUESTIONS AND ANSWERS

—

COMPLETELY REFERENCED

MEDICAL EXAMINATION PUBLISHING COMPANY, INC.
65-36 Fresh Meadow Lane
Flushing, New York 11365

January 1970

Library of Congress
Catalog Card Number
61-66819

Standard Book Number 87488-201-X

PRINTED IN THE UNITED STATES OF AMERICA

PREFACE

There are no short cuts to the process of learning! The material on the following pages is not designed to provide the reader with an opportunity to escape from learning basic information that can only be obtained from textbooks. The purpose of this text is to encourage the reader to detect areas of weakness in his understanding of subject matter so that he may return to his texts for a more comprehensive review of the subject. Crossword puzzles are not designed to teach basic English, examination review books do not teach a basic understanding of the subject. However, the following pages will provide an interesting challenge to the student as well as an opportunity to improve his skills with multiple-choice examinations.

The completely referenced form of the question material is provided to facilitate the reader in checking his areas of weakness. Do not compromise by simply looking up the correct answers — the mature approach is to return to your textbook. The author has provided you with a time saving device in rapidly finding a source of information. While testing your memory be aware of the textbook as being the ultimate source of your fund of knowledge. References cited in the individual questions are listed on page 203.

The variations in test questions are offered to familiarize the reader with various types of National Board, State Board and School examinations in use today. The following test material will give you an opportunity to determine your ability to read, digest and comprehend the vast accumulation of available knowledge. Don't be discouraged if you are unable to achieve a perfect test score; indeed be stimulated to learn the material by referring to your textbooks. Successfully solving the problems in this text will surely give you satisfaction.

SECTION I - THE HEAD AND NECK

FOR EACH OF THE FOLLOWING MULTIPLE CHOICE QUESTIONS SELECT THE ONE MOST APPROPRIATE ANSWER:

1. THE POINT OF MEETING OF THE LAMBDOIDAL, OCCIPITOMASTOID AND PARIETOMASTOID SUTURES IS CALLED THE:
   A. Bregma
   B. Lambda
   C. Asterion
   D. Inion
   E. Pterion — Ref. 3 - p. 50

2. THE FLATTENED TRIANGULAR AREA BETWEEN THE TWO SUPERCILIARY RIDGES IS THE:
   A. Bregma
   B. Glabella
   C. Obelion
   D. Nasion
   E. Inion — Ref. 3 - p. 51

3. PULSATIONS FELT JUST ABOVE THE ZYGOMATIC ARCH AND IN FRONT OF THE EAR ARE PULSATIONS IN THE:
   A. Facial vein
   B. Superficial temporal artery
   C. Lingual artery
   D. Internal carotid artery
   E. External maxillary artery — Ref. 3 - p. 52

4. THE CAROTID OR CHASSAIGNAC'S TUBERCLE IS AN IMPORTANT LANDMARK FOR:
   A. Compressing the common carotid artery to stop bleeding
   B. Locating the site to inject Novocaine in tooth extractions
   C. Making a tracheotomy incision
   D. Locating the recurrent laryngeal nerve
   E. Palpating suboccipital lymph nodes — Ref. 3 - p. 57

5. THE CRICOID CARTILAGE OF THE LARYNX IS AT THE LEVEL OF THE:
   A. 3rd cervical vertebra
   B. 4th cervical vertebra
   C. 6th cervical vertebra
   D. 2nd thoracic vertebra
   E. Hyoid bone — Ref. 3 - p. 57

6. THE MOST PROMINENT MUSCLE IN THE NECK IS THE:
   A. Trapezius
   B. Splenius capitis
   C. Platysma
   D. Sternocleidomastoideus
   E. Digastricus — Ref. 3 - p. 58

SECTION I - THE HEAD AND NECK

7. WHICH ONE OF THE FOLLOWING STRUCTURES IS NOT LOCATED IN THE INFERIOR CAROTID OR MUSCULAR TRIANGLE OF THE NECK:
   A. Sternohyoideus muscle
   B. Isthmus of the thyroid gland
   C. Larynx
   D. Trachea
   E. Branches of the facial nerve Ref. 3 - p. 58

8. THE SUBMANDIBULAR OR DIGASTRIC TRIANGLE OF THE NECK IS BOUNDED ANTERIORLY BY THE:
   A. Mid-line of the body
   B. Anterior belly of the Digastricus
   C. Posterior belly of the Digastricus
   D. Stylohyoideus
   E. Platysma Ref. 3 - p. 59

9. THE EXTERNAL JUGULAR VEIN RUNS A COURSE FROM THE ANGLE OF THE MANDIBLE TO THE:
   A. Suprasternal notch
   B. Mastoid process
   C. Xiphoid process
   D. Middle of the clavicle
   E. Acromion Ref. 3 - p. 61

10. THE MOST DISTINCTIVE FEATURE OF THE SECOND CERVICAL VERTEBRA IS THE:
   A. Odontoid process (the dens)
   B. Transverse process without foramina
   C. Absence of the spinous process
   D. Lack of ligamentous attachments
   E. None of the above Ref. 3 - p. 119

11. THE TEMPORAL FOSSA COMMUNICATES WITH THE ORBITAL CAVITY THROUGH THE:
   A. Inferior orbital or sphenomaxillary fissure
   B. Superior orbital or sphenotemporal fissure
   C. Foramen rotundum
   D. Foramen lacertum
   E. Stephanion Ref. 3 - p. 146

12. THREE FORAMINA OPEN ON THE FACE ON A VERTICAL LINE THAT PASSES BETWEEN THE PREMOLAR TEETH, SUPRA-ORBITAL FORAMEN, INFRA-ORBITAL FORAMEN AND:
   A. Jugular foramen
   B. Foramen lacerum
   C. Mental foramen
   D. Sphenoid foramen
   E. Mastoid foramen Ref. 2 - p. 539

SECTION I - THE HEAD AND NECK

13. WHICH ONE OF THE FOLLOWING IS CALLED THE "GRINNING MUSCLE"?:
A. Masseter
B. Risorius
C. Zygomaticus major
D. Levator anguli oris
E. Platysma Ref. 2 - p. 541

14. WHICH OF THE FOLLOWING MUSCLES PRESSES THE CHEEKS AGAINST THE TEETH:
A. Masseter
B. Buccinator
C. Risorius
D. Temporalis
E. Mentalis Ref. 2 - p. 542

15. THE FACIAL ARTERY (EXTERNAL MAXILLARY) IS CROSSED SUPERFICIALLY BY ALL OF THE FOLLOWING MUSCLES, EXCEPT THE:
A. Platysma
B. Risorius
C. Zygomaticus major
D. Levator labii superioris
E. Buccinator Ref. 2 - p. 549

16. WHICH OF THE FOLLOWING IS THE NERVE TO THE STERNOMASTOID AND TRAPEZIUS MUSCLES?:
A. Accessory nerve
B. Vertebral nerve
C. Occipital nerve
D. Greater occipital nerve
E. Suboccipital nerve Ref. 2 - p. 553

17. FOUND ALONG THE FREE EDGE OF THE FALX CEREBRI IS THE:
A. Superior sagittal sinus
B. Inferior sagittal sinus
C. Great cerebral vein of Galen
D. Straight sinus
E. Transverse sinus Ref. 2 - p. 574

18. ALL VENOUS SINUSES OF THE DURA MATER ULTIMATELY DRAIN INTO THE:
A. External jugular veins
B. Internal jugular veins
C. Emissary veins
D. Carotid sinus
E. Vein of Galen Ref. 2 - p. 576

19. THE LARGEST OF THE CRANIAL NERVES IS THE:
A. Facial
B. Oculomotor
C. Trigeminal
D. Glossopharyngeal
E. Optic Ref. 2 - p. 579

SECTION I - THE HEAD AND NECK

20. THE VERTEBRAL ARTERY PASSES FORWARD TO THE LOWER BORDER OF THE PONS WHERE IT UNITES WITH ITS COUNTERPART TO FORM THE:
A. Basilar artery
B. Internal carotid artery
C. Middle meningeal artery
D. Hypoglossal artery
E. None of the above Ref. 2 - p. 580

21. THE UPWARD EXTENSION OF THE NASAL SEPTUM IS CALLED THE:
A. Galea aponeurotica
B. Lambda
C. Inion
D. Cribiform plate
E. Crista galli Ref. 2 - p. 582

22. INSIDE BOTH LAYERS OF DURA MATER, IN THE MIDDLE CRANIAL FOSSA, ARE ALL OF THE FOLLOWING, EXCEPT FOR THE:
A. Optic chiasma
B. Optic nerve
C. Hypophysis cerebri
D. Temporal lobe of cerebrum
E. Middle meningeal vessels Ref. 2 - p. 587

23. THE CRANIAL NERVE WHICH ARISES ON THE BACK OF THE BRAIN AND THEREFORE HAS THE LONGEST INTRACRANIAL COURSE IS THE:
A. Olfactory
B. Optic
C. Trochlear
D. Oculomotor
E. Trigeminal Ref. 2 - p. 589

24. DOUBLE VISION ON TRYING TO LOOK DOWNWARD OCCURS WITH PARALYSIS OF THE:
A. Abducens nerve
B. Trochlear nerve
C. Oculomotor nerve
D. All of the above
E. None of the above Ref. 2 - p. 589

25. THE FACIAL ARTERY ENTERS THE FACE BY CROSSING THE MANDIBLE AT THE ANTERIOR BORDER OF THE:
A. Masseter muscle
B. Buccinator muscle
C. Platysma muscle
D. Submaxillary gland
E. Parotid gland Ref. 5 - p. 144

SECTION I - THE HEAD AND NECK

26. THE LESSER OCCIPITAL NERVE CHARACTERISTICALLY ASCENDS IN THE NECK ALONG THE:
A. Anterior border of the trapezius muscle
B. Posterior border of the trapezius muscle
C. Anterior border of the sternocleidomastoid muscle
D. Posterior border of the sternocleidomastoid muscle
E. Inferior belly of the digastric muscle
Ref. 5 - p. 148

27. THE SHEATH OF THE THYROID GLAND IS DERIVED FROM THE:
A. Buccopharyngeal fascia
B. Pretracheal fascia
C. Carotid sheath
D. Sibson's fascia
E. Prevertebral fascia — Ref. 5 - p. 152

28. ALL OF THE FOLLOWING ARE FOUND IN THE CAROTID SHEATH, EXCEPT FOR THE:
A. Common carotid artery
B. Internal carotid artery
C. Internal jugular vein
D. Vagus nerve
E. Superior lymphatic duct — Ref. 5 - p. 153

29. THE RETROPHARYNGEAL SPACE IS BOUNDED POSTERIORLY BY THE:
A. Prevertebral fascia
B. Buccopharyngeal fascia
C. Carotid sheath
D. Platysma muscle
E. All of the above — Ref. 5 - p. 154

30. THE INFERIOR BELLY OF THE OMOHYOID MUSCLE ARISES FROM THE:
A. Sternum
B. Hyoid bone
C. Clavicle
D. Mandible
E. None of the above — Ref. 5 - p. 154

31. THE LINGUAL ARTERY ARISES FROM THE:
A. External carotid
B. Submaxillary
C. Internal carotid
D. Common carotid
E. Mandibular — Ref. 2 - p. 682

32. THE INFERIOR THYROID ARTERY DOES NOT SUPPLY THE:
A. Thyroid gland
B. Submandibular gland
C. Parathyroid glands
D. Larynx
E. Pharynx — Ref. 2 - p. 621

SECTION I - THE HEAD AND NECK

33. THE ZYGOMATIC BONE SERVES AS A POINT OF ORIGIN FOR THE:
    A. Frontalis
    B. Temporalis
    C. Buccinator
    D. Masseter
    E. Platysma — Ref. 2 - p. 629

34. A CERVICAL VERTEBRA CAN BE DIFFERENTIATED FROM A LUMBAR VERTEBRA BY THE PRESENCE OF A:
    A. Foramen in each transverse process
    B. Bifid spinous process
    C. Narrow vertebral canal
    D. Cartilagenous body
    E. Bifid vertebral body — Ref. 3 - p. 118

35. THE CHIEF PECULIARITY OF THE FIRST CERVICAL VERTEBRA IS THAT IT HAS:
    A. Bifid spinous processes
    B. No foramen transversarium
    C. No body
    D. No transverse processes
    E. No posterior tubercle — Ref. 3 - p. 118

36. THE POINT OF JUNCTION OF THE SAGITTAL AND CORONAL SUTURES IS CALLED THE:
    A. Corona
    B. Bregma
    C. Lambda
    D. Inion
    E. Amnion — Ref. 3 - p. 145

37. PASSING THROUGH THE GREATER PALATINE FORAMEN ARE THE:
    A. Anterior palatine nerves
    B. Descending palatine arteries
    C. Descending palatine veins
    D. None of the above
    E. All of the above — Ref. 3 - p. 153

38. IMMEDIATELY BEHIND THE CAROTID CANAL IS THE:
    A. Petrosphenoidal suture
    B. Foramen magnum
    C. Medial nuchal line
    D. Jugular foramen
    E. Styloid process — Ref. 3 - p. 158

39. WHICH ONE OF THE FOLLOWING BONES ENTERS INTO THE FORMATION OF THE FLOOR OF THE NASAL CAVITY:
    A. Lacrimal bone
    B. Palatine bone
    C. Ethmoid bone
    D. Temporal bone
    E. Frontal bone — Ref. 3 - p. 201

SECTION I - THE HEAD AND NECK

40. OSSIFIED IN THE FIBROUS MEMBRANE COVERING THE OUTER SURFACE OF MECKEL'S CARTILAGE:
A. Maxilla
B. Mandible
C. Frontal
D. Occipital
E. Ethmoid
Ref. 3 - p. 166

41. THE CORONOID PROCESS IS PART OF THE ______________ BONE:
A. Parietal
B. Mandible
C. Sphenoid
D. Temporal
E. Ethmoid
Ref. 3 - p. 165

42. THE HYOID BONE IS OSSIFIED FROM ______________ CENTERS:
A. 2
B. 3
C. 4
D. 5
E. 6
Ref. 3 - p. 170

43. WHICH ONE OF THE FOLLOWING NERVES PASSES THROUGH THE FORAMEN MAGNUM:
A. Spinal
B. Vagus
C. Accessory
D. Hypoglossal
E. All of the above
Ref. 3 - p. 172

44. THE GLABELLA IS FOUND ON THE ______________ BONE:
A. Sphenoid
B. Occipital
C. Frontal
D. Parietal
E. Temporal
Ref. 3 - p. 176

45. WHICH OF THE FOLLOWING BONES IS BEST DESCRIBED AS A "BAT WITH WINGS EXTENDED"?:
A. Mandible
B. Occipital
C. Lacrimal
D. Sphenoid
E. Ethmoid
Ref. 3 - p. 186

46. FORAMINA ROTUNDUM, OVALE AND SPINOSUM ARE SITUATED IN THE ______________ BONE:
A. Frontal
B. Ethmoid
C. Maxillary
D. Temporal
E. None of the above
Ref. 3 - p. 180

SECTION I - THE HEAD AND NECK

47. THE UNCINATE PROCESS IS PART OF THE ____________ BONE:
A. Sphenoid
B. Temporal
C. Ethmoid
D. Frontal
E. Lacrimal Ref. 3 - p. 193

48. THE MAXILLARY SINUS, OR ANTRUM OF HIGHMORE, COMMUNICATES WITH THE ____________ OF THE NOSE:
A. Superior meatus
B. Inferior meatus
C. Middle meatus
D. Vestibule
E. Posterior nares Ref. 3 - pp. 197

49. THE PROCERUS MUSCLE IS ASSOCIATED WITH THE:
A. Mouth
B. Nose
C. Ear
D. Tongue
E. Scalp Ref. 3 - p. 386

50. THE STERNOCLEIDOMASTOID MUSCLE IS INNERVATED BY THE ____________ NERVE:
A. Vagus
B. Accessory
C. Hypoglossal
D. Brachial
E. Trigeminal Ref. 3 - p. 400

51. THE GENIOHYOIDEUS IS INNERVATED BY THE ____________ NERVE:
A. Facial
B. Mandibular
C. Accessory
D. Vagus
E. None of the above Ref. 3 - p. 401

52. BRANCHES OF THE EXTERNAL CAROTID ARTERY INCLUDE THE:
A. Superficial temporal
B. Internal maxillary
C. Ascending pharyngeal
D. Occipital
E. All of the above Ref. 3 - p. 584

53. WHICH OF THE FOLLOWING VESSELS TAKES ORIGIN FROM THE EXTERNAL CAROTID ARTERY AT A LEVEL JUST BELOW THE GREATER CORNU OF THE HYOID BONE?:
A. Superior thyroid
B. Superior laryngeal
C. External maxillary
D. Posterior auricular
E. Internal maxillary Ref. 3 - p. 584

SECTION I - THE HEAD AND NECK

54. THE LARGEST ARTERY SUPPLYING THE DURA MATER OF THE BRAIN IS THE:
A. Anterior cerebral
B. Middle meningeal
C. Anterior meningeal
D. Posterior meningeal
E. Artery of the pterygoid canal — Ref. 3 - p. 592

55. BRANCHES OF THE CERVICAL PORTION OF THE INTERNAL CAROTID ARTERY INCLUDE THE:
A. Caroticotympanic
B. Artery of the pterygoid canal
C. Cavernous
D. Anterior meningeal
E. None of the above — Ref. 3 - p. 596

56. THE BASILAR ARTERY TERMINATES BY DIVIDING INTO TWO ARTERIES:
A. Anterior inferior cerebellar arteries
B. Superior cerebellar arteries
C. Posterior cerebral arteries
D. Internal auditory arteries
E. Anterior communicating arteries — Ref. 3 - p. 607

57. THE SMALLEST OF THE CRANIAL SINUSES IS THE:
A. Superior sagittal
B. Inferior sagittal
C. Straight
D. Occipital
E. Transverse — Ref. 3 - p. 695

58. THE CAVERNOUS SINUS RECEIVES THE SUPERIOR OPHTHALMIC VEIN VIA THE:
A. Superior orbital fissure
B. Inferior orbital fissure
C. Optic foramen
D. Infraorbital foramen
E. Sphenopalatine foramen — Ref. 3 - p. 695

59. THE LYMPHATIC VESSELS OF THE TONGUE DRAIN PRIMARILY INTO THE:
A. Palatine nodes
B. Deep cervical nodes
C. Gingival nodes
D. Submaxillary nodes
E. Parotid nodes — Ref. 3 - p. 746

60. THE SMALLEST OF THE CRANIAL NERVES IS THE:
A. Olfactory
B. Oculomotor
C. Trochlear
D. Accessory
E. Abducens — Ref. 3 - p. 916

SECTION I - THE HEAD AND NECK

61. THE CILIARY GANGLION IS LOCATED IN THE:
A. Nasal cavity
B. Ethmoidal sinus
C. Frontal sinus
D. Maxillary sinus
E. None of the above Ref. 3 - p. 914

62. THE NERVUS SPINOSUS ENTERS THE SKULL THROUGH THE:
A. Foramen rotundum
B. Foramen ovale
C. Foramen spinosum
D. Foramen magnum
E. Jugular foramen Ref. 3 - p. 924

63. WHICH ONE OF THE FOLLOWING IS FOUND IN THE MANDIBULAR CANAL?:
A. Inferior alveolar nerve
B. Mylohyoid nerve
C. Mental artery
D. Lingual artery
E. Buccinator ligament Ref. 3 - p. 927

64. THE ZYGOMATICUS MAJOR MUSCLE HAS ITS MAIN ACTION IN:
A. Crying
B. Frowning
C. Laughing
D. Kissing
E. Smoking Ref. 3 - p. 387

65. WHICH ONE OF THE FOLLOWING MUSCLES OF MASTICATION INSERTS INTO THE UPPER HALF OF THE RAMUS AND LATERAL SURFACE OF THE CORONOID PROCESS OF THE MANDIBLE?:
A. Masseter
B. Temporalis
C. Pterygoideus internus
D. Pterygoideus externus
E. All of the above Ref. 3 - p. 393

66. THE PLATYSMA IS INNERVATED BY THE:
A. Mandibular division of V
B. Maxillary division of V
C. XI
D. Cervical branch of VII
E. XII Ref. 3 - p. 394

67. WHICH ONE OF THE FOLLOWING IS REGARDED AS A LATERAL VERTEBRAL MUSCLE?:
A. Longus colli
B. Longus capitis
C. Rectus capitis anterior
D. Rectus capitis posterior
E. Scalenus anterior Ref. 3 - p. 404

SECTION I - THE HEAD AND NECK

68. THE FIRST BRANCH OF THE EXTERNAL CAROTID ARTERY IS USUALLY THE:
A. Lingual
B. Superior laryngeal
C. Superior thyroid
D. Infracricoid
E. Middle cervical   Ref. 5 - p. 157

69. ON ITS WAY TO THE TONGUE, THE LINGUAL ARTERY IS CROSSED BY THE:
A. Facial vein
B. Middle meningeal artery
C. Wharton's duct
D. Hypoglossal nerve
E. All of the above   Ref. 5 - p. 160

70. AN INFECTION IN THE NOSE AND MEDIAL PORTION OF THE EYELID WILL PROBABLY CAUSE ENLARGEMENT OF THE ____________ LYMPH NODES:
A. Superficial parotid
B. Deep cervical
C. Occipital
D. Retroauricular
E. Anterior auricular   Foramen   Ref. 5 - p. 165

71. THE CHORDA TYMPANI, A BRANCH OF THE __________ NERVE, LEAVES THE PETROUS PORTION OF THE TEMPORAL BONE THROUGH THE __________ FISSURE:
A. Lingual, pterygopalatine
B. Facial, petrotympanic
C. Hypoglossal, sphenoethmoidal
D. Lingual, petrotympanic
E. Facial, sphenoethmoidal   Ref. 5 - p. 173

72. THE SUBLINGUAL SALIVARY GLAND RESTS ON THE:
A. Mylohyoid muscle
B. Anterior belly of the digastric muscle
C. Omohyoid muscle
D. Trapezius muscle
E. Hyoid bone   Ref. 5 - p. 174

73. WHICH ONE OF THE FOLLOWING IS AN UNPAIRED MUSCLE OF THE LARYNX:
A. Cricothyroid
B. Posterior cricoarytenoid
C. Lateral cricoarytenoid
D. Transverse arytenoid
E. Thyroarytenoid   Ref. 5 - pp. 187-188

SECTION I - THE HEAD AND NECK

74. THE CORRUGATOR MUSCLE INSERTS INTO THE:
A. Medial part of the superciliary arch
B. Medial half of the clavicle
C. Skin of the corner of the mouth
D. Skin of the medial half of the eyebrow
E. None of the above Ref. 5 - p. 200

75. THE ANGULAR VEIN IS FORMED BY THE UNION OF THE ________ AND ________ VEINS OF THE FOREHEAD:
A. Supratrochlear, supraorbital
B. Supratrochlear, infraorbital
C. External maxillary, facial
D. Facial, external jugular
E. Lacrimal, nasolabial Ref. 5 - p. 203

ANSWER THE FOLLOWING QUESTIONS ACCORDING TO THE KEY BELOW:
A. If both statement and reason are true and related cause and effect
B. If both statement and reason are true but not related cause and effect
C. If the statement is true but the reason is false
D. If the statement is false but the reason is true
E. If both statement and reason are false

76. The facial nerve is a pure motor nerve
because
It provides the motor innervation of the muscles of facial expression.
Ref. 5 - p. 207

77. The mandibular nerve is composed of a motor and sensory root
because
The two roots leave the middle cranial fossa through the foramen ovale.
Ref. 5 - p. 221

78. Swelling of the submandibular lymph nodes denotes cancer of the lower lip
because
Lymphatics of the upper lip do not pass to these lymph nodes.
Ref. 5 - p. 226

79. The growth of the face is approximately four times that of the cranial portion of the head
because
The permanent teeth are larger and more numerous than the deciduous teeth.
Ref. 5 - p. 226

80. The muscles of the two sides of the tongue are separated by the lingual septum
because
The muscles of the tongue are divided into extrinsic and intrinsic groups.
Ref. 5 - p. 229

SECTION I - THE HEAD AND NECK

81. Tears may pass from the orbital region to the nasal cavity
because
The nasolacrimal duct terminates in the superior meatus of the lateral wall of the nose.
Ref. 5 - p. 237

82. The paranasal sinuses are pneumatic areas found in the facial and cranial bones at the second month of fetal life
because
Absorption of bone around the invading nasal mucosal sacs in these bones establishes the paranasal sinuses.
Ref. 5 - p. 238

83. Lymphatics have not been found in the eyeball or orbit
because
The blood supply of the eyelids is quite rich.
Ref. 5 - pp. 242-243

84. Infections and thromboses in the superior and inferior ophthalmic veins are never associated with involvement of the brain
because
Both of these veins have numerous valves.
Ref. 5 - pp. 251-252

85. The three auditory ossicles are united by true joints
because
They are suspended in the cavity of the middle ear.
Ref. 5 - p. 259

86. The body of the mandible is curved like a horseshoe
because
It serves for the reception of the lower teeth.
Ref. 3 - p. 163

87. The stylomandibular ligament is a major ligament of the temporomandibular joint
because
It separates the parotid and submaxillary glands.
Ref. 3 - p. 302

88. The facial or external maxillary artery is quite tortuous
because
It must accommodate for the movements of the mandible, cheeks, and lips.
Ref. 3 - p. 586

89. The facial artery is called an end artery
because
It does not anastomose with other vessels.
Ref. 3 - p. 586

SECTION I - THE HEAD AND NECK

90. The middle cerebral artery is the largest branch of the internal carotid
because
It supplies the frontal, parietal, temporal, and occipital lobes, as well as the cerebellum.
Ref. 3 - p. 602

91. An infection in the ala nasi is always considered a serious medical problem
because
Its venous pathway leads directly to the sagittal sinus of the brain.
Ref. 3 - p. 693

92. Enlarged superior cervical lymph nodes may suggest an infection of the palatine tonsils
because
Lymphatics from this structure pass to the uppermost of the deep cervical nodes.
Ref. 3 - p. 746

93. Palpable superficial cervical lymph nodes suggest carcinoma of the posterior margin of the tongue
because
The lymphatics from this area of the tongue pass first to the submaxillary nodes then to the superficial cervical nodes.
Ref. 3 - p. 746

94. The cervical plexus is situated opposite the upper four cervical vertebrae
because
It is formed by the dorsal primary divisions of the upper four cervical nerves.
Ref. 3 - pp. 959-960

95. The larynx contains only intrinsic ligaments
because
The intrinsic ligaments connect the cartilages of the larynx to the hyoid bone.
Ref. 3 - p. 1129

FOR EACH OF THE FOLLOWING QUESTIONS IDENTIFY THE UNRELATED OR INCORRECT ITEM:

96. ARTERIES OF THE MIDDLE EAR:
A. Anterior tympanic branch of the maxillary artery
B. Stylomastoid branch of the posterior auricular artery
C. Inferior tympanic branch of the ascending pharyngeal artery
D. Superficial petrosal branch of the middle meningeal artery
E. Superficial ossicle branch of the internal carotid artery
Ref. 5 - p. 259

SECTION I - THE HEAD AND NECK

97. VENOUS SINUSES OF THE DURA MATER:
    A. Inferior sagittal sinus
    B. Cavernous sinus
    C. Great cerebral sinus
    D. Sigmoid sinus
    E. Transverse sinus — Ref. 5 - pp. 274-275

98. BRANCHES OF THE VERTEBRAL ARTERY:
    A. Anterior spinal
    B. Meningeal
    C. Posterior spinal
    D. Middle cerebral
    E. Posterior inferior cerebellar — Ref. 5 - p. 277

99. PARASYMPATHETIC GANGLIA IN THE HEAD:
    A. Sphenoethmoidal
    B. Ciliary
    C. Pterygopalatine
    D. Submandibular
    E. Otic — Ref. 5 - p. 281

100. INFRAHYOID MUSCLES:
    A. Sternothyroid
    B. Mandibulohyoid
    C. Thyrohyoid
    D. Omohyoid
    E. Sternohyoid — Ref. 5 - p. 154

101. BRANCHES OF THE EXTERNAL CAROTID ARTERY ARISING IN THE CAROTID TRIANGLE:
    A. Superior thyroid
    B. Superficial temporal
    C. Lingual
    D. Facial
    E. Ascending pharyngeal — Ref. 5 - p. 160

102. TRIBUTARIES OF THE INTERNAL JUGULAR VEIN:
    A. Vein of Galen
    B. Inferior petrosal sinus
    C. Lingual
    D. Superior thyroid
    E. Facial — Ref. 5 - p. 162

103. SUPERFICIAL LYMPH NODES OF THE HEAD:
    A. Occipital
    B. Retroauricular
    C. Anterior auricular
    D. Superficial parotid
    E. Anterior jugular — Ref. 5 - p. 165

SECTION I - THE HEAD AND NECK

104 BRANCHES OF THE SUBCLAVIAN ARTERY:
A. Vertebral
B. Thyrocervical
C. Occipital
D. Internal thoracic
E. Dorsal scapular — Ref. 5 - p. 179

105. CARTILAGES OF THE LARYNX:
A. Thyroid
B. Arytenoid
C. Cuneiform
D. Corniculate
E. Pisiform — Ref. 5 - pp. 183-184

106. MUSCLES OF THE PHARYNX:
A. Stylopharyngeus
B. Palatopharyngeus
C. Mandibulopharyngeus
D. Salpingopharyngeus
E. Inferior constrictor — Ref. 5 - p. 192

107. MUSCLES OF THE MOUTH:
A. Corrugator
B. Orbicularis oris
C. Zygomaticus major
D. Buccinator
E. Risorius — Ref. 5 - pp. 200-201

108. PAIRED BONES OF THE SKULL:
A. Frontal
B. Nasal
C. Maxilla
D. Zygomatic
E. Temporal — Ref. 4 - p. 161

109. MUSCLES ATTACHED TO THE MANDIBLE:
A. Masseter
B. Temporalis
C. Platysma
D. Triangularis
E. Caninus — Ref. 4 - pp. 223-225

110. BONES FORMING THE EXTERNAL BASE OF THE SKULL:
A. Maxilla
B. Zygomatic
C. Parietal
D. Temporal
E. Sphenoid — Ref. 4 - p. 168

SECTION I - THE HEAD AND NECK

111. MUSCLES INSERTING ON THE OCCIPITAL BONE:
A. Rectus capitis posterior minor
B. Semispinalis capitis
C. Splenius capitis
D. Trapezius
E. Sternocleidomastoideus — Ref. 4 - p. 184

112. SUTURES INCLUDED IN THE MIDDLE CRANIAL FOSSA:
A. Sphenoparietal
B. Sphenosquamosal
C. Squamous
D. Sphenopetrosal
E. Parietal — Ref. 4 - p. 175

113. PARTS OF THE FRONTAL BONE:
A. Glabella
B. Superciliary arch
C. Metopic suture
D. Mastoid angle
E. Frontal spine — Ref. 4 - p. 190-192

114. SUPRAHYOID MUSCLES:
A. Stylohyoid
B. Digastric
C. Mylohyoid
D. Geniohyoid
E. Masseter — Ref. 4 - p. 452

115. BRANCHES OF THE EXTERNAL CAROTID ARTERY:
A. Superior laryngeal
B. Superior thyroid
C. Lingual
D. Occipital
E. Ascending pharyngeal — Ref. 4 - p. 674

116. BRANCHES OF THE SUPERFICIAL TEMPORAL ARTERY:
A. Anterior auricular
B. Posterior auricular
C. Parotid
D. Transverse facial
E. Parietal — Ref. 4 - p. 682

117. BRANCHES OF THE FIRST PART OF THE INTERNAL MAXILLARY ARTERY:
A. Deep auricular
B. Anterior tympanic
C. Lingual
D. Middle meningeal
E. Inferior alveolar — Ref. 4 - pp. 683-684

SECTION I - THE HEAD AND NECK

118. BRANCHES OF THE INTRACRANIAL PORTION OF THE INTERNAL CAROTID ARTERY:
A. Ophthalmic
B. Posterior communicating
C. Choroid
D. Anterior cerebral
E. Middle meningeal Ref. 4 - p. 689

119. BRANCHES OF THE OPHTHALMIC ARTERY:
A. Lacrimal
B. Posterior ciliary
C. Anterior ethmoidal
D. Dorsal nasal
E. Superior labial Ref. 4 - p. 689

120. TRIBUTARIES OF THE RETROMANDIBULAR VEIN:
A. Lacrimal
B. Superficial temporal
C. Temporomandibular
D. Transverse facial
E. Maxillary Ref. 4 - p. 794

121. TRIBUTARIES OF THE EXTERNAL JUGULAR VEIN:
A. Internal maxillary
B. Posterior auricular
C. Occipital
D. Transverse scapular
E. Anterior jugular Ref. 4 - p. 795

122. UNPAIRED VENOUS SINUSES OF THE DURA MATER:
A. Superior sagittal
B. Transverse
C. Straight
D. Occipital
E. Circular Ref. 4 - p. 798

123. DEEP VEINS OF THE NECK:
A. Internal jugular
B. Vertebral
C. Anterior jugular
D. Inferior thyroid
E. Esophageal Ref. 4 - pp. 807-809

124. PAIRED SINUSES OF THE BRAIN:
A. Occipital
B. Transverse
C. Sigmoid
D. Inferior petrosal
E. Cavernous Ref. 4 - p. 798

SECTION I - THE HEAD AND NECK

125. AREAS DRAINED BY THE LYMPHATIC SUBMENTAL NODES:
A. Skin of chin
B. Floor of mouth
C. Tip of tongue
D. Maxillary sinus
E. Central part of lower lip Ref. 4 - p. 875

126. BRANCHES OF THE OPHTHALMIC NERVE:
A. Frontal
B. Supraorbital
C. Supratrochlear
D. Lacrimal
E. Zygomatic Ref. 4 - pp. 1029-1030

127. BRANCHES OF THE MANDIBULAR NERVE:
A. Spinous
B. Superior labial
C. Buccal
D. Pterygoideus medialis
E. Lingual Ref. 4 - p. 1032

128. NERVES COMING FROM THE SUPERIOR CERVICAL GANGLION:
A. Internal carotid
B. Jugular
C. Laryngopharyngeal
D. Masseteric
E. External carotid Ref. 4 - pp. 1105-1107

129. EFFECTS OF PARALYSIS OF THE CERVICAL SYMPATHETIC TRUNK IN THE NECK:
A. Narrowing of the pupil
B. Ptosis
C. Enophthalmos
D. Loss of dilator functioning of pupil
E. Excess lacrimation Ref. 4 - p. 1105

130. PARTS OF THE EXTERNAL EAR:
A. Helix
B. Schrapnell's membrane
C. Cymba conchae
D. Scapha
E. Tragus Ref. 4 - pp. 1175-1176

131. ARTERIES OF THE TYMPANIC CAVITY:
A. Superior tympanic from occipital
B. Anterior tympanic from internal maxillary
C. Stylomastoid from posterior auricular
D. Superficial petrosal from middle meningeal
E. Inferior tympanic from ascending pharyngeal
Ref. 4 - p. 1193

SECTION I - THE HEAD AND NECK

132. EXTRINSIC MUSCLES OF THE TONGUE:
A. Hyoglossus
B. Chondroglossus
C. Glossopalatinus
D. Genioglossus
E. Styloglossus Ref. 4 - p. 1239

133. ANATOMICAL LANDMARKS OF THE MIDDLE NASAL MEATUS:
A. Frontal recess
B. Semilunar hiatus
C. Uncinate process
D. Ostium of nasolacrimal duct
E. Ethmoidal bulla Ref. 4 - p. 1395

134. EXTRINSIC LARYNGEAL MUSCLES:
A. Omohyoid
B. Thyrohyoid
C. Sternohyoid
D. Palatopharyngeus
E. Cricothyroid Ref. 4 - p. 1418

135. PARANASAL SINUSES:
A. Frontal
B. Parietal
C. Ethmoidal
D. Maxillary
E. Sphenoidal Ref. 4 - p. 1397

FOLLOWING EACH GROUP OF FIVE CHOICES THERE ARE EITHER FIVE STATEMENTS OR FIVE ITEMS. MATCH THE STATEMENTS OR ITEMS WITH THE MOST APPROPRIATE CHOICE GIVEN. EACH CHOICE MAY BE USED ONLY ONCE:

A. Pháryngopalatinus muscle
B. Inferior pharyngeal constrictor
C. Middle pharyngeal constrictor
D. Stylopharyngeus muscle
E. Levator veli palatini muscle

136. ___ Arises from thyroid cartilage, cranial fibers insert into fibrous raphe of pharynx, constricts pharynx

137. ___ Inserts into aponeurosis of soft palate, in part arises from thyroid cartilage, innervated by vagus, one action is to depress soft palate

138. ___ Fan-shaped muscle, in part arises from lesser cornu of hyoid bone, nerve supply is the vagus

139. ___ Originates in part from inferior surface of petrous portion of temporal bone, inserts into soft palate, one action is to close pharyngeal isthmus

140. ___ Inserts in part into dorsal border of thyroid cartilage, innervated by glossopharyngeal nerve, draws pharyngeal wall cranialward
Ref. 4 - pp. 1272-1277

SECTION I - THE HEAD AND NECK

A. Ophthalmic artery
B. Central artery of retina
C. Lacrimal artery
D. Supraorbital artery
E. Short posterior ciliary artery

141. ___ Arises close to optic canal and pierces optic nerve, passes forward in center of optic nerve

142. ___ Branch of intracranial portion of internal carotid artery

143. ___ Terminal distribution is by two lateral palpebral arteries

144. ___ Distributes to skin and musculature of forehead, anastomoses with supratrochlear artery

145. ___ Pierces sclera around optic nerve entrance, supplies choroid and ciliary processes

Ref. 5 - pp. 250-251

A. Superior oblique muscle
B. Levator palpebrae superioris muscle
C. Lateral rectus muscle
D. Medial rectus muscle
E. Inferior oblique muscle

146. ___ Innervated by oculomotor nerve, about 40 mm. long, inserts into sclera

147. ___ Innervated by abducens nerve, slender and flat, inserts into sclera

148. ___ Innervated by oculomotor nerve, originates at back of orbit in front of optic foramen, inserts along upper border of tarsus

149. ___ Shortest of all voluntary ocular muscles, takes origin from medial margin of floor of orbit, innervated by oculomotor nerve

150. ___ Longest and thinnest eye muscle, arises from periosteum of sphenoid bone, innervated by trochlear nerve Ref. 4 - pp. 1154-1159

A. Maxillary nerve
B. Nerve of pterygoid canal
C. Accessory meningeal artery
D. Ophthalmic artery
E. Inferior petrosal sinus

151. ___ Foramen ovale

152. ___ Foramen lacerum

153. ___ Foramen rotundum

154. ___ Jugular foramen

155. ___ Optic foramen

Ref. 3 - pp. 153,158

SECTION I - THE HEAD AND NECK

FOLLOWING THE CASE HISTORY ARE SEVERAL QUESTIONS BASED ON THE HISTORY. CHOOSE THE BEST ANSWER FOR EACH QUESTION:

A four year-old boy has a chronic history of earaches and tonsillitis. On one occasion he has had a draining right ear and x-rays revealed a right mastoid infection. The child required prolonged and intensive antibiotic therapy. At the present time he has a moderate hearing loss and fluid is seen behind both tympanic membranes. A decision is made to perform a tonsillectomy and adenoidectomy. The patient is seen by an otolaryngologist and the patient scheduled for surgery. A tonsillectomy involving the palatine tonsils precludes a knowledge of the anatomy of the area.

156. THE CAPSULE OF THE PALATINE TONSIL LIES AGAINST THE:
A. Superior pharyngeal constrictor muscle
B. Palatopharyngeus muscle
C. Styloglossus muscle
D. All of the above
E. None of the above Ref. 4 - p. 1266

157. A TONSILLAR ABSCESS MAY EXTEND THROUGH THE PHARYNGEAL WALL INTO THE:
A. Pleural cavity
B. Anterior triangle of the neck
C. Suprasternal space
D. Waldeyer's ring
E. Retropharyngeal space Ref. 4 - p. 1266

158. THE PALATINE TONSIL IS SEPARATED, BY THE PHARYNGEAL WALL, FROM THE:
A. Ascending palatine artery
B. External carotid artery
C. Internal carotid artery
D. Lingual artery
E. Medial pterygoid muscle Ref. 4 - p. 1266

159. FOLLOWING SURGERY A PULSATING ARTERY IS NOTED AND REQUIRES LIGATION; IT WAS PROBABLY THE:
A. Posterior tonsillar from the pterygoid
B. Anterior tonsillar from the dorsal lingual
C. Inferior tonsillar from the dorsal facial
D. Dorsal tonsillar from the posterior pharyngeal
E. Superior tonsillar from the inferior lingual
Ref. 4 - p. 1266

160. IF SURGERY COULD HAVE BEEN POSTPONED, INVOLUTION OF THE TONSIL WOULD PROBABLY HAVE BEGUN AT AGE ___________ YEARS:
A. 5
B. 7
C. 13
D. 18
E. 21 Ref. 4 - p. 1267

SECTION I - THE HEAD AND NECK

IDENTIFY EACH OF THE CRANIAL NERVES OR THEIR PARTS, AS INDICATED BY THE GUIDE LINES.

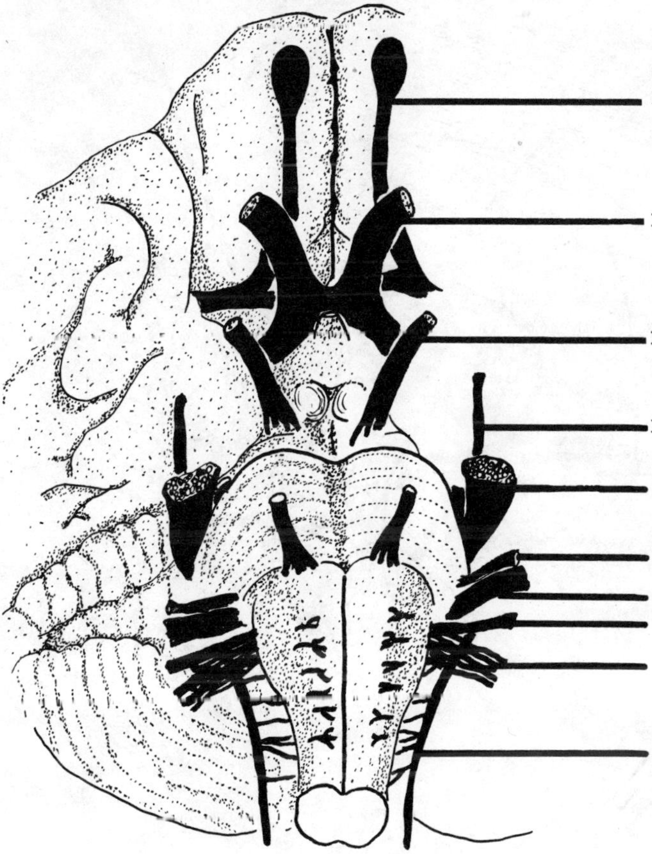

A. Optic nerve
B. Vagus nerve
C. Trochlear nerve
D. Optic tract
E. Olfactory tract
F. Abducens nerve
G. Spinal nerve
H. Semilunar ganglion
I. Glossopharyngeal nerve
J. Auditory nerve
K. Maxillary nerve
L. Facial nerve
M. Accessory nerve
N. Oculomotor nerve
O. Trigeminal nerve

Ref. 1 - p. 506

SECTION I - THE HEAD AND NECK

IDENTIFY THE MUSCLES ORIGINATING OR INSERTING ON THE SKULL IN THE AREAS INDICATED BY THE GUIDE LINES.

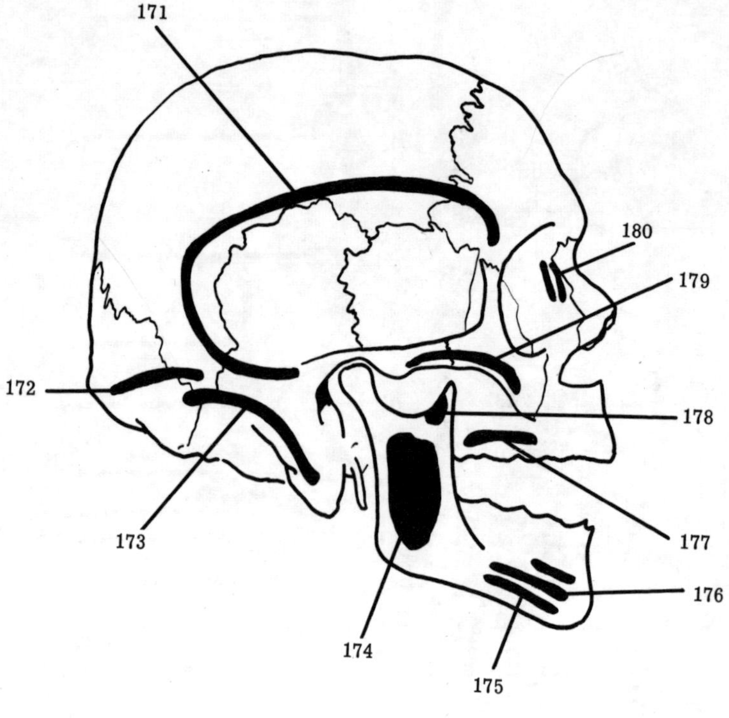

A. Buccinator
B. Temporalis
C. Quadratus labii superioris
D. Occipitalis
E. Styloglossus
F. Platysma
G. Zygomaticus
H. Thyroglossus
I. Triangularis
J. Omohyoideus
K. Sternocleidomastoideus
L. Labialis
M. Orbicularis oris
N. Masseter

Ref. 4 - p. 167

SECTION I - THE HEAD AND NECK

IDENTIFY THE ARTERIES FORMING THE BLOOD SUPPLY OF THE BRAIN (CIRCLE OF WILLIS).

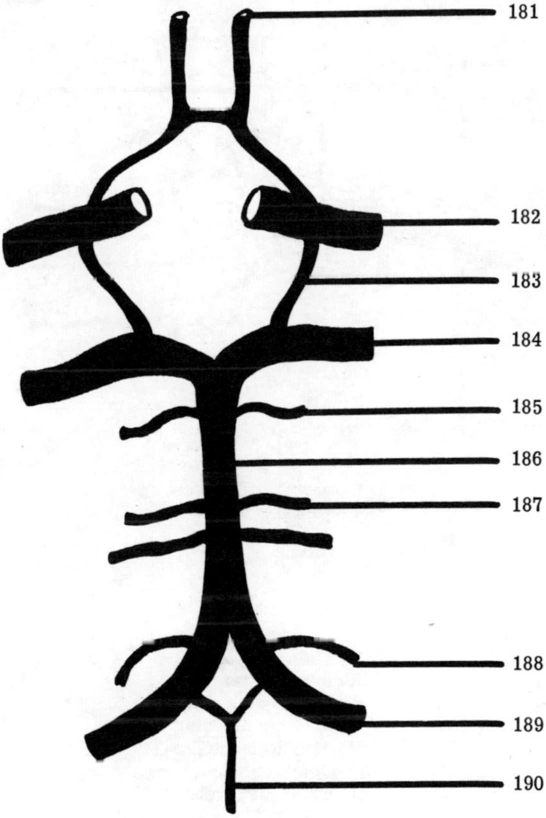

A. Anterior cerebral
B. Middle cerebral
C. Posterior communicating
D. Posterior cerebral
E. Superior cerebellar
F. Basilar
G. Internal auditory
H. Posterior inferior cerebellar
I. Vertebral
J. Anterior spinal
K. Anterior inferior cerebellar
L. External carotid
M. Anterior communicating
N. Middle cerebellar
O. Lateral cerebellar

Ref. 2 - p. 580

SECTION I - THE HEAD AND NECK

IDENTIFY THE BRANCHES OF THE INTERNAL MAXILLARY ARTERY AS SCHEMATICALLY SHOWN IN THE DIAGRAM BELOW.

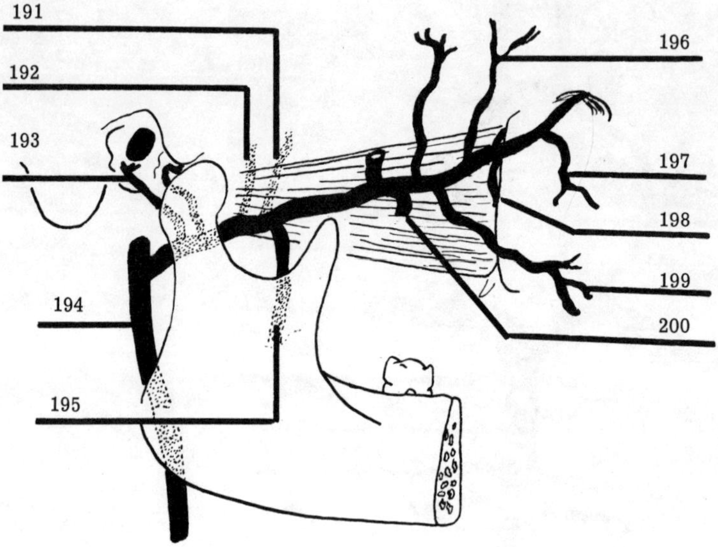

A. Internal carotid
B. Pterygoid
C. Deep auricular
D. Buccal
E. Accessory meningeal
F. Posterior superior alveolar
G. Anterior tympanic
H. Spheno-palatine
I. Deep temporal
J. Infraorbital
K. Descending palatine
L. Inferior alveolar
M. Middle meningeal
N. Common carotid
O. External carotid

Ref. 1 - p. 556

SECTION I - THE HEAD AND NECK

IDENTIFY THE STRUCTURES ON THE LATERAL ASPECT OF THE SKULL.

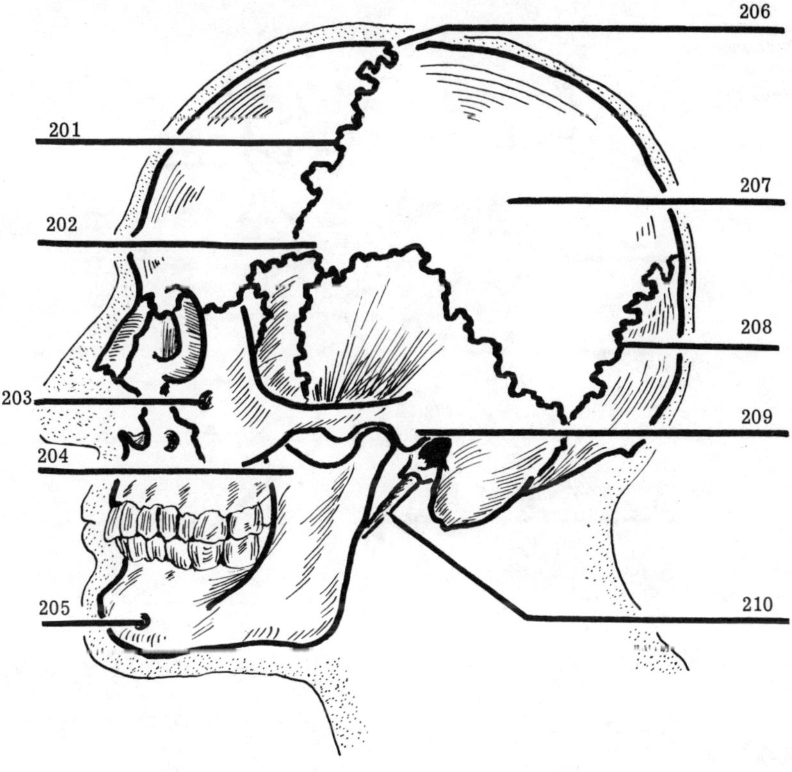

A. Pterion
B. Coronoid process
C. Coronal suture
D. Infraorbital foramen
E. Parietal bone
F. Lambdoid suture
G. Lambda
H. Tubercle
I. Styloid process
J. Postglenoid tubercle
K. Temporal bone
L. Zygomatico-facial foramen
M. Mental foramen
N. Condyloid process
O. Bregma

Ref. 1 - pp. 461-462

SECTION I - THE HEAD AND NECK

IDENTIFY THE MUSCLES, TRIANGLES, AND ASSOCIATED STRUCTURES OF THE NECK AS INDICATED BY THE GUIDE LINES.

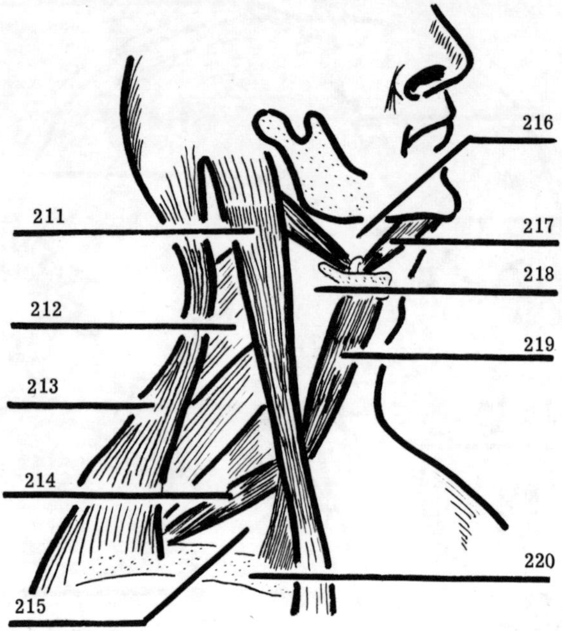

A. Sternomastoid
B. Splenius
C. Trapezius
D. Clavicle
E. Scapula
F. Occipital triangle
G. Levator scapulae
H. Inferior belly of Omohyoid
I. Subclavian triangle
J. Digastric triangle
K. Scalenus anterior
L. Thyroid gland
M. Anterior belly of Digastric
N. Carotid triangle
O. Superior belly of Omohyoid

Ref. 1 - pp. 540-543

SECTION I - THE HEAD AND NECK

IDENTIFY THE SENSORY NERVES OF THE FACE AS INDICATED BY THE GUIDE LINES.

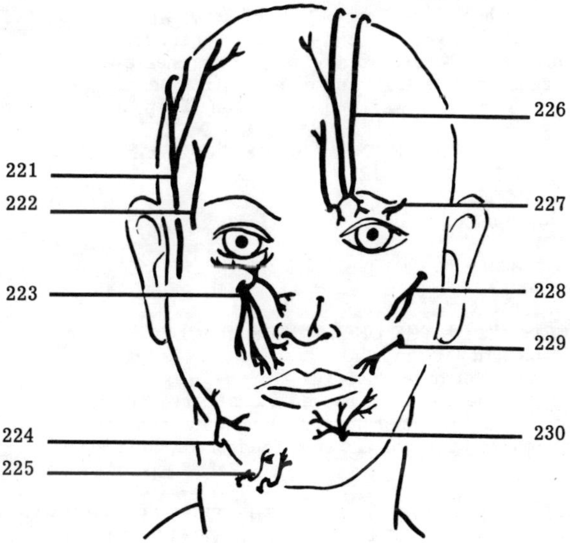

A. Infraorbital
B. Transverse cervical
C. Zygomatico-facial
D. Mental
E. Auriculo-temporal
F. Supraorbital
G. Great auricular
H. Zygomatico-temporal
I. Lacrimal
J. Buccal
K. Infratrochlear
L. External nasal

Ref. 1 - pp. 470-471

SECTION II - THE UPPER EXTREMITY

FOR EACH OF THE FOLLOWING MULTIPLE CHOICE QUESTIONS SELECT THE APPROPRIATE ANSWER OR ANSWERS. ONE OR MORE, OR ALL OF THE CHOICES MAY BE USED IN CORRECTLY ANSWERING THE QUESTION.

231. NERVES TO THE TRAPEZIUS MUSCLE ARE THE:
A. Accessory nerve (cranial nerve XI)
B. Branch of the hypoglossal nerve (cranial nerve XII)
C. Direct branches of ventral rami of 3rd cervical nerve
D. Direct branches of dorsal rami of 4th cervical nerve
E. Direct branches of ventral rami of 5th cervical nerve
Ref. 5 - p. 51

232. MUSCLES FORMING THE BOUNDARIES OF THE TRIANGLE OF AUSCULTATION INCLUDE THE:
A. Latissimus dorsi
B. Rhomboideus major
C. Trapezius
D. Triceps brachii
E. Subscapularis
Ref. 5 - p. 52

233. CHARACTERISTICS OF THE DELTOIDEUS MUSCLE:
A. Initiates abduction of the humerus
B. Principal abductor of the humerus
C. Innervated by the accessory nerve (cranial nerve XI)
D. Originates from the humerus, clavicle and sternum
E. Lies deep to the trapezius and sternocleidomastoid muscles
Ref. 5 - p. 54

234. BRANCHES OF THE SECOND PART OF THE AXILLARY ARTERY (LYING BEHIND THE TENDON OF THE PECTORALIS MINOR):
A. Highest thoracic
B. Subscapular
C. Anterior circumflex humeral
D. Thoracoacromial
E. Lateral thoracic
Ref. 5 - pp. 78-79

235. THE MEDIAL BRACHIAL CUTANEOUS NERVE:
A. Arises from the medial cord of the brachial plexus
B. Contains fibers from the 8th cervical and 1st thoracic nerves
C. Descends through the axilla behind the axillary vein
D. Pierces the brachial fascia about mid-arm
E. Innverates the coracobrachialis muscle and skin over this muscle
Ref. 5 - p. 82

236. THE BICIPITAL APONEUROSIS PASSES OBLIQUELY OVER THE:
A. Brachial artery
B. Radial nerve
C. Long thoracic nerve
D. Cephalic vein
E. Median nerve
Ref. 5 - p. 87

SECTION II - THE UPPER EXTREMITY

237. THE RADIAL NERVE INNERVATES THE:
A. Triceps
B. Anconeus
C. Coracobrachialis
D. Brachialis
E. Serratus anterior Ref. 5 - p. 89

238. IN MOVEMENTS OF THE ARM, THE CORACOBRACHIALIS ASSISTS IN:
A. Abduction
B. Extension
C. Rotation
D. Supination
E. Flexion Ref. 5 - p. 89

239. POINTS OF ANATOMICAL INTEREST FOUND ON THE HUMERUS:
A. Deltoid tuberosity
B. Radial fossa
C. Trochlea
D. Capitulum
E. Surgical neck Ref. 5 - p. 89

240. THE ULNAR NERVE:
A. Is the largest nerve derived from the medial cord of the brachial plexus
B. Carries fibers from 6C, 7C and 8C
C. In the arm lies lateral to the brachial artery
D. Has no branches in the arm except articular twigs at the elbow
E. Has no sensory function Ref. 5 - p. 91

241. THE BRACHIAL ARTERY:
A. Bifurcates opposite the neck of the humerus
B. At first lies medial to the humerus
C. Is deep throughout its entire course
D. In the cubital fossa is lateral to the biceps tendon
E. Terminates in radial and ulnar arteries
Ref. 5 - p. 93

242. THE CUBITAL FOSSA IS BOUNDED BY THE:
A. Triceps
B. Biceps
C. Coracobrachialis
D. Brachioradialis
E. Pronator quadratus Ref. 5 - p. 96

243. MUSCLES ROTATING THE RADIUS ON THE ULNA:
A. Pronator teres
B. Flexor carpi ulnaris
C. Flexor carpi radialis
D. Palmaris longus
E. Extensor digitorum Ref. 5 - p. 97

SECTION II - THE UPPER EXTREMITY

244. PRE-AXIAL MUSCLES INCLUDE THE:
A. Pronator teres
B. Pronator quadratus
C. Palmaris longus
D. Flexor pollicis longus
E. Supinator Ref. 5 - p. 97

245. SUPERFICIAL GROUP OF POSTERIOR ANTEBRACHIAL MUSCLES INCLUDES:
A. Coracobrachialis
B. Triceps
C. Brachioradialis
D. Extensor carpi ulnaris
E. Supinator Ref. 5 - p. 98

246. POINTS OF ANATOMICAL INTEREST FOUND ON THE ULNA:
A. Olecranon
B. Coronoid process
C. Styloid process
D. Head
E. Supinator crest Ref. 5 - p. 105

247. BRANCHES OF THE ULNAR ARTERY IN THE FOREARM:
A. Anterior ulnar recurrent
B. Common interosseous
C. Muscular
D. Palmar carpal
E. Dorsal carpal Ref. 5 - p. 110

248. THE CARPAL BONES, IN THE PROXIMAL ROW, CONSIST OF:
A. Scaphoid
B. Lunate
C. Triangular
D. Pisiform
E. Trapezium Ref. 5 - pp. 111-112

249. THE PALMAR CARPAL LIGAMENT CROSSES THE:
A. Tendon of palmaris longus
B. Tendon of biceps brachii
C. Ulnar nerve
D. Radial nerve
E. Median nerve Ref. 5 - p. 112

250. THE TRAPEZIUM ARTICULATES WITH THE:
A. Trapezoid
B. Pisiform
C. Radius
D. Scaphoid
E. Lunate Ref. 5 - p. 115

SECTION II - THE UPPER EXTREMITY

251. MUSCLES FORMING THE POSTERIOR WALL OF THE AXILLA:
A. Subscapularis
B. Teres major
C. Latissimus dorsi
D. Pectoralis major
E. Pectoralis minor — Ref. 2 - p. 78

252. CONCERNING THE BRACHIAL PLEXUS:
A. 6 branches
B. 5 trunks
C. 6 divisions
D. 8 rami
E. 3 cords — Ref. 2 - p. 85

253. THE ARTERIAL SUPPLY OF THE MAMMARY GLAND IS DERIVED FROM THE:
A. Superior epigastric
B. Brachial
C. Internal thoracic
D. Intercostals
E. Lateral thoracic — Ref. 2 - pp. 89-90

254. ACTIONS AND FUNCTIONS OF THE LATISSIMUS DORSI MUSCLE:
A. Flexes humerus
B. Laterally rotates humerus
C. Abducts humerus
D. All of the above
E. None of the above — Ref. 2 - p. 94

255. THE SUPRASCAPULAR NERVE SUPPLIES THE:
A. Supraspinatus muscle
B. Infraspinatus muscle
C. Subscapularis muscle
D. Deltoid muscle
E. Trapezius muscle — Ref. 2 - p. 96

256. FORMING BOUNDARIES OF THE QUADRANGULAR SPACE:
A. Rhomboideus major
B. Lateral border of the scapula
C. Biceps tendon
D. Surgical neck of the humerus
E. Teres major — Ref. 2 - pp. 99-100

257. MUSCLES ATTACHED TO THE INFERIOR ANGLE OF THE SCAPULA:
A. Serratus anterior
B. Teres major
C. Rhomboideus major
D. Deltoideus
E. Triceps — Ref. 2 - p. 102

SECTION II - THE UPPER EXTREMITY

258. THE MUSCLES OF THE UPPER LIMB ARE SUPPLIED BY NERVE SEGMENTS:
A. C-5
B. C-6
C. C-7
D. C-8
E. T-1 Ref. 2 - p. 109

259. THE LATERAL CUTANEOUS NERVE OF THE FOREARM IS DERIVED FROM:
A. C-4
B. C-5
C. C-6
D. C-7
E. C-8 Ref. 2 - p. 113

260. THE PALMAR DIGITAL NERVES ARE DERIVED FROM THE:
A. Median nerve
B. Ulnar nerve
C. Radial nerve
D. Lateral cutaneous nerve
E. Musculocutaneous nerve Ref. 2 - p. 113

261. NERVES SUPPLYING THE DORSUM OF THE HAND:
A. Radial
B. Ulnar
C. Median
D. Musculocutaneous
E. Axillary Ref. 2 - p. 113

262. THE CEPHALIC VEIN:
A. Crosses the anatomical "snuffbox"
B. Ascends on the ulnar border of the forearm
C. Pierces the delto-pectoral triangle
D. Ends in the basilic vein
E. May cross in front of the clavicle and ends in the external jugular vein Ref. 2 - p. 115

263. MUSCLES INCLUDED IN THE POSTERIOR COMPARTMENT OF THE ARM:
A. Coracobrachialis
B. Biceps brachii
C. Brachialis
D. Triceps brachii
E. Pectoralis major Ref. 2 - p. 118

264. MUSCLES INNERVATED BY THE MUSCULOCUTANEOUS NERVE:
A. Coracobrachialis
B. Brachialis
C. Triceps brachii
D. All of the above
E. None of the above Ref. 2 - p. 121

SECTION II - THE UPPER EXTREMITY

265. USUAL NUMBER OF BRANCHES OF THE MEDIAN NERVE IN THE ARM:
A. 0
B. 1
C. 3
D. 4
E. 6 Ref. 2 - p. 121

266. AMONG THE CONTENTS OF THE CUBITAL FOSSA:
A. Brachial artery
B. Radial nerve
C. Ulnar nerve
D. Median nerve
E. Radial artery Ref. 2 - p. 124

267. DEEP MUSCULATURE OF THE SHOULDER GIRDLE:
A. Pectoralis major
B. Latissimus dorsi
C. Levator scapulae
D. Deltoideus
E. Serratus anterior Ref. 4 - p. 465

268. INNERVATED BY THE AXILLARY NERVE:
A. Flexor carpi radialis
B. Palmaris longus
C. Pronator teres
D. Teres minor
E. Latissimus dorsi Ref. 4 - p. 472

269. BRANCHES OF THE FIRST PART OF THE LEFT SUBCLAVIAN ARTERY INCLUDE:
A. Costocervical trunk
B. Vertebral
C. Internal mammary
D. Thyrocervical trunk
E. Transverse cervical Ref. 4 - p. 694

270. BRANCHES OF THE RADIAL ARTERY AT THE WRIST:
A. 1st dorsal metacarpal
B. 2nd dorsal metacarpal
C. 3rd dorsal metacarpal
D. All of the above
E. None of the above Ref. 4 - p. 721

271. CONCERNING THE BASILIC VEIN:
A. Arises on dorsum of hand
B. Always superficial to brachial fascia
C. Unites with cephalic to form axillary
D. All of the above
E. None of the above Ref. 4 - p. 812

SECTION II - THE UPPER EXTREMITY

272. INCLUDED AMONG THE DEEP VEINS OF THE UPPER EXTREMITY:
A. Cephalic
B. Basilic
C. Axillary
D. Subclavian
E. Brachial vena comitantes — Ref. 4 - pp. 812-813

273. FEATURES OF THE ULNAR NERVE:
A. From lateral cord of brachial plexus
B. Fibers from C-8 and T-1
C. Muscular branches in upper arm
D. All of the above
E. None of the above — Ref. 4 - p. 1063

274. CONCERNING THE MEDIAN NERVE:
A. Supplies muscles of upper arm
B. Trunk formed below lower margin of pectoralis minor
C. In forearm accompanies median artery
D. All of the above
E. None of the above — Ref. 4 - pp. 1064-1065

275. DEEP LYMPHATIC VESSELS OF THE UPPER EXTREMITY DRAIN THE:
A. Joint capsules
B. Periosteum
C. Tendons
D. Nerves
E. Muscles — Ref. 4 - p. 883

276. THE SUPERFICIAL CUBITAL NODES RECEIVE LYMPH VESSELS FROM THE:
A. Ulnar digits
B. Palm of the hand
C. Thumb
D. Forefinger
E. All of the above — Ref. 4 - p. 883

277. THE EARLIEST BONE TO OSSIFY IS THE:
A. Radius
B. Ulna
C. Clavicle
D. Humerus
E. Mandible — Ref. 3 - p. 210

278. THE CONSTRICTION BELOW THE GREATER AND LESSER TUBERCLES OF THE HUMERUS IS CALLED THE:
A. Anatomical neck
B. Surgical neck
C. Humeral shaft
D. Constriction ring
E. None of the above — Ref. 3 - p. 216

SECTION II - THE UPPER EXTREMITY

279. MUSCLES ASSOCIATED WITH THE GREATER TUBERCLE OF THE HUMERUS INCLUDE THE:
A. Supraspinatus
B. Infraspinatus
C. Teres major
D. Teres minor
E. Subscapularis Ref. 3 - p. 215

280. LANDMARKS ASSOCIATED WITH THE DISTAL END OF THE HUMERUS:
A. Surgical neck
B. Greater tubercle
C. Trochlea
D. Capitulum
E. None of the above Ref. 3 - p. 216

ANSWER THE FOLLOWING QUESTIONS ACCORDING TO THE KEY BELOW:
A. If only A is correct
B. If only B is correct
C. If A and B are correct
D. If A and B are incorrect

281. THE CLAVICLE:
A. Is subcutaneous in its entire length
B. Is the most frequently fractured bone of the body
C. Both
D. Neither Ref. 5 - pp. 59-74

282. THE PECTORAL FASCIA:
A. Invests the subclavius and pectoralis minor muscles
B. Is continuous with the fascia of the teres major
C. Both
D. Neither Ref. 5 - p. 76

283. BRANCHES OF THE THIRD PART OF THE AXILLARY ARTERY:
A. Anterior circumflex humeral
B. Lateral thoracic
C. Both
D. Neither Ref. 5 - p. 78

284. THE DORSAL SCAPULAR NERVE:
A. Is the nerve to the levator scapulae muscle
B. Arises from the posterior aspect of the ventral ramus of C-5
C. Both
D. Neither

285. THE SUBSCAPULARIS MUSCLE:
A. Fills the dorsal surface of the scapula
B. Is the chief medial rotator of the arm
C. Both
D. Neither
Ref. 5 - p. 85

286. THE CORACOBRACHIALIS MUSCLE:
A. Inserts into the tip of the coracoid process
B. Is pierced at midlength by the median nerve
C. Both
D. Neither Ref. 5 - p. 86

SECTION II - THE UPPER EXTREMITY

287. THE BICIPITAL APONEUROSIS:
A. Passes obliquely over the brachial artery and ulnar nerve
B. Blends with the antebrachial fascia over the extensor muscles of the forearm
C. Both
D. Neither Ref. 5 - p. 87

288. PRINCIPAL FLEXORS OF THE FOREARM AT THE ELBOW:
A. Brachialis
B. Biceps brachii
C. Both
D. Neither Ref. 5 - p.89

289. THE MEDIAN NERVE:
A. Is the largest nerve derived from the medial cord of the brachial plexus
B. Crosses the brachial artery at mid-arm
C. Both
D. Neither Ref. 5 - p. 91

290. THE BRACHIAL ARTERY:
A. Extends from the lower border of the teres major muscle
B to the styloid process of the radius
B. Is deep throughout its entire course
C. Both
D. Neither Ref. 5 - p. 93

291. THE CHARACTERISTIC "ROUNDNESS" OF THE SHOULDER IS DUE TO THE:
A. Acromion process of the scapula
B. Coracoid process of the scapula
C. Both
D. Neither Ref. 4 - pp. 248-249

292. THE FURROW BETWEEN THE GREATER AND LESSER TUBERCLES OF THE HUMERUS LODGES THE:
A. Axillary artery
B. Long tendon of the biceps
C. Both
D. Neither Ref. 4 - p. 250

293. ATTACHED TO THE STYLOID PROCESS OF THE ULNA:
A. Ulnar collateral ligament of the wrist joint
B. Pronator teres muscle
C. Both
D. Neither Ref. 4 - p. 261

294. ACTIONS OF THE EXTENSOR CARPI ULNARIS MUSCLE:
A. Abduct the hand ulnarward
B. Extend the 5th metacarpal
C. Both
D. Neither Ref. 4 - p. 495

SECTION II - THE UPPER EXTREMITY

295. MUSCLES INNERVATED BY THE MEDIAN NERVE:
A. Flexor pollicis longus
B. Flexor digitorum profundus
C. Both
D. Neither Ref. - pp. 504-506

296. MUSCLES SITUATED SUPERFICIAL TO THE THENAR FASCIA:
A. Abductor pollicis brevis
B. Opponens pollicis
C. Both
D. Neither Ref. 4 - pp. 514-516

297. ANASTOMOSES OF BRANCHES OF THE BRACHIAL ARTERY:
A. Volar ulnar recurrent and superior ulnar collateral
B. Radial recurrent and radial collateral
C. Both
D. Neither Ref. 4 - p. 714

298. BRANCHES OF THE ULNAR ARTERY IN THE FOREARM:
A. Common interosseous
B. Superficial volar
C. Both
D. Neither Ref. 4 - p. 715

299. ARTERIAL TRIBUTARIES OF THE DEEP VOLAR ARCH:
A. Princeps pollicis
B. Volaris indicis radialis
C. Both
D. Neither Ref. 4 - p. 722

300. CONCERNING THE AXILLARY VEIN:
A. Formed by fusion of cephalic and basilic veins
B. Empties into base of inferior vena cava
C. Both
D. Neither Ref. 4 - p. 014

ANSWER THE FOLLOWING QUESTIONS USING THE KEY AS OUTLINED BELOW:
A. If 1, 2 and 3 are correct
B. If 1 and 3 are correct
C. If 2 and 4 are correct
D. If all are correct
E. If some combination other than the above are correct
F. If all are incorrect

301. GUARDING THE LATERAL BOUNDARY OF THE FLEXOR REGION OF THE FOREARM:
1. Insertion of the brachialis muscle
2. Insertion of the biceps muscle
3. Insertion of the pronator teres muscle
4. Insertion of the brachioradialis muscle
Ref. 2 - p. 128

SECTION II - THE UPPER EXTREMITY

302. THE FLEXOR DIGITORUM SUBLIMIS ARISES FROM:
1. Anterior oblique line of radius between attachments of flexor pollicis longus and supinator
2. Anterior fibrous cord of ulnar collateral ligament
3. Intramuscular septa
4. Antebrachial fascia of forearm Ref. 2 - p. 130

303. CARPAL BONES IN THE DISTAL ROW:
1. Scaphoid
2. Lunate
3. Triquetrum
4. Pisiform Ref. 2 - p. 137

304. THE PISIFORM ARTICULATES WITH THE:
1. Triquetrum
2. Lunate
3. Scaphoid
4. Capitate Ref. 2 - p. 138

305. THE SECOND METACARPAL ARTICULATES WITH THE:
1. Trapezium
2. Lunate
3. Capitate
4. Hamate Ref. 2 - p. 141

306. OSSIFICATION OF THE CARPAL BONES:
1. Capitate, first year
2. Triquetrum, third year
3. Lunate, fourth year
4. Pisiform, twelfth year Ref. 2 - p. 143

307. MUSCLES INNERVATED BY THE MEDIAN NERVE:
1. Palmaris brevis
2. Opponens pollicis
3. Adductor pollicis
4. 1st lumbrical Ref. 2 - p. 149

308. STRUCTURES CROSSING THE SKIN CREASE AT THE WRIST AND ENTERING THE PALM; SUPERFICIAL TO THE CARPAL TUNNEL:
1. Flexor pollicis longus muscle
2. Ulnar nerve
3. Median nerve
4. Superficial palmar branch of radial artery

Ref. 2 - p. 151

309. MUSCLES PRODUCING A DEPRESSION OF THE SHOULDER GIRDLE WITH DOWNWARD ROTATION OF THE GLENOID CAVITY:
1. Rhomboids
2. Pectoralis minor
3. Trapezius (mid.)
4. Serratus anterior Ref. 2 - p. 175

SECTION II - THE UPPER EXTREMITY

310. MUSCLE ACTING ON THE RADIO-ULNAR JOINTS:
1. Pronator quadratus
2. Pronator teres
3. Flexor carpi radialis
4. Biceps Ref. 2 - p. 189

311. MUSCLES ORIGINATING ON THE CLAVICLE:
1. Subclavius
2. Trapezius
3. Pectoralis major
4. Deltoideus Ref. 3 - p. 209

312. MUSCLES INSERTING ON THE SCAPULA:
1. Serratus anterior
2. Trapezius
3. Pectoralis minor
4. Deltoideus Ref. 3 - pp. 211-212

313. ANATOMICAL FEATURES OF THE HUMERUS:
1. Greater tubercle lateral to the head
2. Lesser tubercle more prominent than the greater tubercle
3. Anatomical neck provides attachment for articular capsule
4. Fracture of anatomical neck rarely occurs
Ref. 3 - pp. 215-216

314. MUSCLES ORIGINATING FROM THE RADIUS:
1. Biceps
2. Supinator
3. Pronator quadratus
4. Anconeus Ref. 3 - pp. 221, 225

315. ANATOMICAL FEATURES OF THE TRAPEZIUS MUSCLE:
1. In part originates from the occipital bone
2. In part inserts into the clavicle
3. Upper part acting alone draws the scapula upward
4. Innervated by the axillary nerve Ref. 3 - p. 450

316. MUSCLES ASSOCIATED WITH THE "TRIANGLE OF AUSCULTATION":
1. Trapezius
2. Pectoralis major
3. Latissimus dorsi
4. Pectoralis minor Ref. 3 - p. 452

317. THE AXILLARY FASCIA IS CONTINUOUS WITH THE:
1. Pectoral fascia
2. Brachial fascia
3. Prevertebral fascia
4. Pretracheal fascia Ref. 3 - p. 455

318. CONCERNING THE PECTORALIS MAJOR MUSCLE:
1. Arises from the humerus
2. Inserts into the ribs
3. Abducts the arm
4. Innervated by the median nerve Ref. 3 - p. 456

SECTION II - THE UPPER EXTREMITY

319. MUSCLES OF THE SHOULDER JOINT INCLUDE:
1. Subclavius
2. Subscapularis
3. Supraspinatus
4. Infraspinatus Ref. 3 - p. 458

320. THE BRACHIALIS MUSCLE:
1. Covers the anterior aspect of the elbow joint
2. Flexes the forearm
3. Inserts into the tuberosity of the ulna and the anterior surface of the coronoid process
4. Is innervated by the ulnar nerve Ref. 3 - p. 463

321. AMONG THE SUPERFICIAL GROUP OF PALMAR ANTEBRACHIAL MUSCLES:
1. Flexor digitorum profundus
2. Pronator teres
3. Pronator quadratus
4. Flexor carpi ulnaris Ref. 3 - p. 465

322. BRANCHES OF THE BRACHIAL ARTERY:
1. Profunda brachii
2. Radial collateral
3. Inferior ulnar collateral
4. Interosseous recurrent Ref. 3 - p. 620

323. RELATIONS OF THE ULNAR ARTERY AT THE WRIST:
1. Lies on the flexor retinaculum
2. Pisiform bone on its medial side
3. Ulnar nerve dorsal to artery
4. Covered by palmar carpal ligament Ref. 3 - p. 625

324. SUPERFICIAL VEINS OF THE UPPER EXTREMITY:
1. Do not accompany arteries
2. Communicate with the deep veins
3. Are larger than the deep veins
4. Do not have valves Ref. 4 - p. 809

325. THE BASILIC VEIN:
1. Arises on the dorsum of the hand
2. Joins the internal jugular vein
3. Accompanies the median antebrachial cutaneous nerve in the forearm
4. Is a deep vein Ref. 4 - p. 812

326. TRIBUTARIES OF THE AXILLARY VEIN:
1. Thoracic epigastric
2. Costoaxillary
3. Cephalic
4. Right brachiocephalic Ref. 4 - p. 814

SECTION II - THE UPPER EXTREMITY

327. THE BRACHIAL PLEXUS IS FORMED IN PART BY THE VENTRAL PRIMARY ARM OF:
1. C-5
2. C-6
3. C-7
4. C-8 Ref. 4 - p. 1059

328. THE RADIAL NERVE:
1. Is the smallest branch of the posterior cord of the brachial plexus
2. Contains fibers from C-1, C-2, C-3, and C-4
3. In the arm lies ventral to the brachial artery and posterior to the long head of the triceps
4. Terminates at mid-forearm Ref. 4 - pp. 1066-1067

329. THE NERVE SUPPLY TO THE ELBOW JOINT:
1. Musculocutaneous
2. Ulnar
3. Median
4. Radial Ref. 4 p. 363

330. THE NERVE SUPPLY TO THE WRIST ARTICULATION:
1. Deep branch of the ulnar
2. Anterior interosseous of the median
3. Radial recurrent of the musculocutaneous
4. Posterior interosseous of the median Ref. 4 - p. 369

FOLLOWING THE CASE HISTORY ARE SEVERAL QUESTIONS BASED ON THE HISTORY. CHOOSE THE BEST ANSWER FOR EACH QUESTION.

A 20 year-old college student is injured in an automobile accident. In the Emergency Room of a nearby hospital he is examined and found to have multiple abrasions and lacerations. In addition, he is found to have a fracture through the upper half of the right arm. The patient is admitted to the hospital and more thorough examination reveals that the ulnar nerve has been severed. As a life-saving measure it was necessary for the physician in the Emergency Room to ligate the brachial artery in the upper-third of the arm. However, attempts at suturing the cut ends of the ulnar nerve were unsuccessful.

331. AS A RESULT OF THE INJURY TO THE ULNAR NERVE AT THE WRIST:
A. Power of flexion is diminished
B. Power of extension is diminished
C. Ulnar adduction is lost
D. Ulnar abduction is lost
E. A and C are correct
F. B and D are correct
G. All are correct Ref. 4 - pp. 1063-1064

SECTION II - THE UPPER EXTREMITY

332. CHANGES IN THE HAND WILL INCLUDE:
A. Marked wasting of the hypothenar eminence
B. Marked wasting of the thenar eminence
C. The thumb cannot be abducted
D. The thumb cannot be adducted
E. A and C are correct
F. B and D are correct
G. All are correct Ref. 4 - pp. 1063-1064

333. LOSS OF SENSATION WILL BE DETECTED OVER THE:
A. Thumb
B. 2nd and 3rd digits
C. 4th and 5th digits
D. A, B and C are correct
E. None of the above are correct Ref. 4 - p. 1065

334. IN LIGATING THE BRACHIAL ARTERY AS DESCRIBED:
A. No collateral circulation can be established
B. Collateral circulation is possible only if ligation is below the level of the superior ulnar collateral artery
C. Collateral circulation is possible if ligation is above the level of the superior ulnar collateral
D. Immediate amputation is necessary
E. Only amputation of the fingers is necessary
Ref. 4 - p. 714

335. COLLATERAL CIRCULATION MAY BE ESTABLISHED AT THIS LEVEL BY:
A. Superior ulnar collateral and inferior ulnar collateral
B. Radial collateral and middle collateral
C. Anterior circumflex and posterior circumflex
D. All of the above
E. None of the above Ref. 4 - p. 714

SECTION II - THE UPPER EXTREMITY

FOLLOWING EACH GROUP OF FIVE CHOICES THERE ARE EITHER FIVE STATEMENTS OR FIVE ITEMS. MATCH THE STATEMENTS OR ITEMS WITH THE MOST APPROPRIATE CHOICE GIVEN. EACH CHOICE MAY BE USED ONLY ONCE:

A. Latissimus dorsi
B. Trapezius
C. Teres major
D. Deltoideus
E. Serratus anterior

336. ___ Triangular in shape; suspensory muscle of shoulder girdle

337. ___ Forms medial wall of axilla; inserts into scapula

338. ___ Extends the humerus, rotates the humerus medially; used in swimming and rowing

339. ___ Arises from anterior surface of clavicle

340. ___ Passes between the axillary and radial nerves

Ref. 2 - pp. 94-95, 97, 99, 102-103

A. Musculo-cutaneous nerve
B. Ulnar nerve
C. Median nerve
D. Radial nerve
E. Posterior interosseous nerve

341. ___ No branches in the arm; two roots unite a little below the Pectoralis minor

342. ___ Arises from medial cord of brachial plexus; during its course applied to medial head of the Triceps

343. ___ Terminal branch of posterior cord; during course lies on capsule of elbow joint

344. ___ Essentially motor; disappears into the substance of the Supinator

345. ___ Arises from lateral cord of plexus; pierces deep fascia as lateral cutaneous nerve of forearm

Ref. 2 - pp. 121-122

SECTION II - THE UPPER EXTREMITY

A. "Anatomical snuff-box"
B. Carpal tunnel
C. Formation of osseo-fascial tunnels for tendons
D. Proximal radio-ulnar joint
E. Crosses skin crease superficially to enter palm

346. ___ Annular ligament

347. ___ Ulnar nerve

348. ___ Radial artery

349. ___ Flexor retinaculum

350. ___ Extensor retinaculum

Ref. 2 - pp. 133-134, 139, 146, 151, 161

SECTION II - THE UPPER EXTREMITY

IDENTIFY THE STRUCTURES INDICATED IN THE SCHEMATIC CROSS SECTION THROUGH THE WRIST.

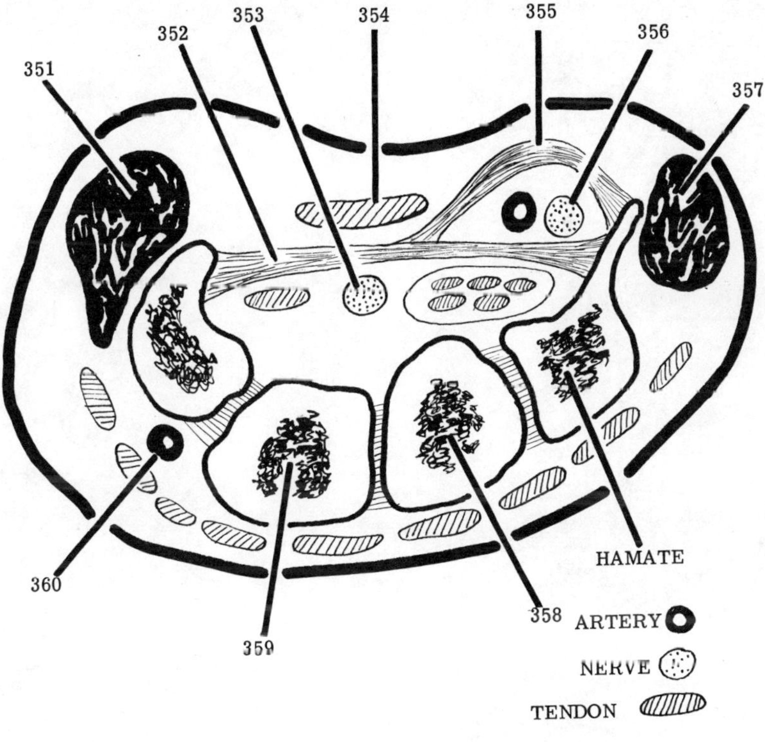

A. Median nerve
B. Tendon of Palmaris longus
C. Capitate bone
D. Tendon of Flexor carpi radialis
E. Radial artery
F. Pisiform bone
G. Greater multangular bone
H. Ulnar artery
I. Extensor retinaculum
J. Ulnar nerve
K. Short muscles of little finger
L. Flexor retinaculum
M. Radial nerve
N. Volar carpal ligament
O. Short muscles of thumb

Ref. 1 - pp. 64-72
Ref. 2 - p. 146

SECTION II - THE UPPER EXTREMITY

ON THE SCHEMATIC OUTLINE OF THE ANTERIOR VIEW OF THE BONES OF THE FOREARM AND HAND, IDENTIFY THE ORIGINS AND INSERTIONS OF THE VARIOUS MUSCLES AS INDICATED BY THE GUIDE LINES.

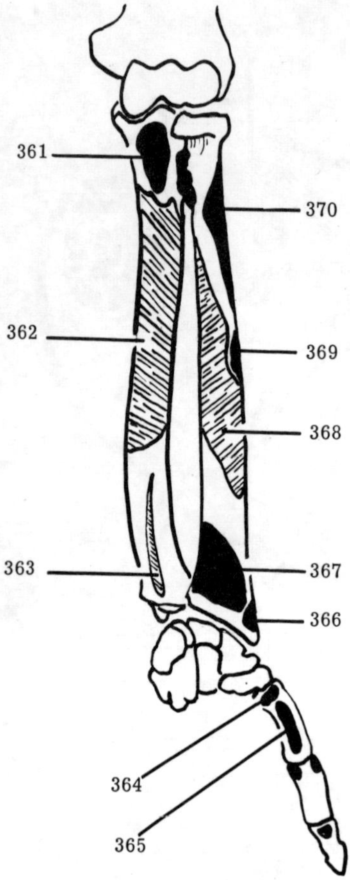

A. Pronator quadratus
B. Abductor pollicis longus
C. Pronator teres
D. Interossei
E. Brachialis
F. Flexor pollicis longus
G. Brachioradialis
H. Supinator
I. Flexor digitorum profundus
J. Opponens pollicis
K. Flexor carpi radialis
L. Flexor digitorum superficialis
M. Abductor flexor brevis
N. Extensor carpi ulnaris
O. Biceps brachii

Ref. 1 - p. 58

SECTION II - THE UPPER EXTREMITY

ON THE SCHEMATIC OUTLINE OF THE ANTERIOR SURFACE OF THE HUMERUS, IDENTIFY THE ORIGINS AND INSERTIONS OF THE VARIOUS MUSCLES AS INDICATED BY THE GUIDE LINES.

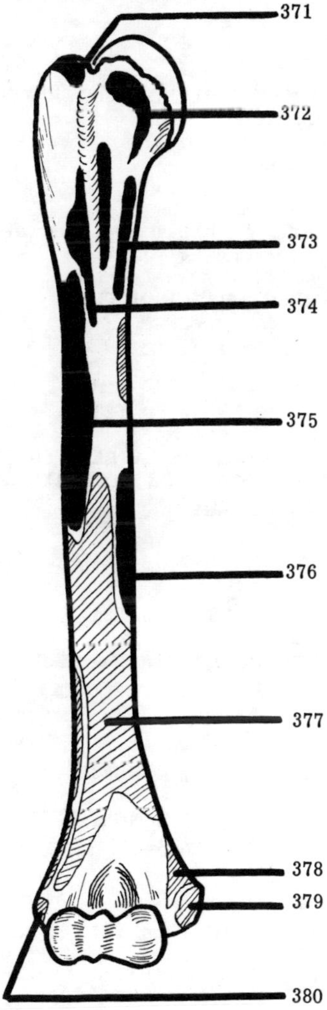

A. Teres major
B. Common extensor tendon
C. Biceps brachii
D. Pronator teres
E. Subscapularis
F. Supraspinatus
G. Coracobrachialis
H. Common flexor tendon
I. Brachialis
J. Triceps (long head)
K. Latissimus dorsi
L. Pectoralis minor
M. Pectoralis major
N. Trapezius
O. Deltoideus

ORIGIN ////////
INSERTION ▬

Ref. 1 - p. 29

SECTION III - THE LOWER EXTREMITY

FOR EACH OF THE FOLLOWING MULTIPLE CHOICE QUESTIONS SELECT THE ONE APPROPRIATE ANSWER.

381. THE GREATER SAPHENOUS VEIN:
A. Is the longest vein of the body
B. Lies next to the anterior femoral cutaneous nerve in the thigh
C. Empties into the femoral vein
D. All of the above
E. None of the above Ref. 5 - p. 519

382. ALL OF THE FOLLOWING ARE CHARACTERISTIC OF THE FEMORAL BRANCH OF THE GENITOFEMORAL NERVE, EXCEPT:
A. Arises from the lumbar plexus
B. Supplies the skin over the femoral triangle
C. Pierces the femoral sheath
D. Enters the thigh in front of the inguinal ligament
E. Communicates with the anterior cutaneous branch of the femoral nerve
Ref. 5 - p. 522

383. THE SAPHENOUS NERVE IS A TERMINAL BRANCH OF THE:
A. Ilioinguinal nerve
B. Femoral nerve
C. Genitofemoral nerve
D. Superior cluneal nerve
E. Popliteal nerve Ref. 5 - p. 523

384. THE ____________ NERVE IS CUTANEOUS TO THE LATERAL SIDE AND THE BACK OF THE DISTAL ONE-THIRD OF THE LEG AND TO THE ANKLE AND THE HEEL:
A. Superficial peroneal
B. Greater saphenous
C. Sural
D. Lateral femoral
E. Medial calcaneal Ref. 5 - p. 524

385. THE ILIOTIBIAL TRACT RECEIVES TENDINOUS REINFORCEMENT FROM:
A. Tensor fasciae latae
B. Gluteal aponeurosis
C. Inguinal ligament
D. Sacrotuberous ligament
E. None of the above Ref. 5 - p. 526

386. THE LATERAL BOUNDARY OF THE FEMORAL TRIANGLE IS THE:
A. Sartorius muscle
B. Inguinal ligament
C. Adductor longus muscle
D. Iliopsoas muscle
E. Gluteal aponeurosis Ref. 5 - p. 527

387. THE FEMORAL SHEATH IS A DIVERTICULUM OF THE:
A. Gluteal aponeurosis
B. Iliotibial tract
C. Transversalis fascia
D. Medial intermuscular septum
E. Fascia cribrosa Ref. 5 - p. 527-528

SECTION III - THE LOWER EXTREMITY

388. WHICH ONE OF THE FOLLOWING IS NOT FOUND IN THE ADDUCTOR CANAL:
A. Saphenous nerve
B. Femoral artery
C. Femoral vein
D. Lower nerve to vastus medialis
E. Saphenous vein Ref. 5 - p. 528

389. ALL OF THE FOLLOWING ARE POSTAXIAL MUSCLES OF THE LATERAL FEMORAL GROUP, EXCEPT FOR THE:
A. Gluteus maximus
B. Gluteus minimus
C. Piriformis
D. Obturator internus
E. Tensor fasciae latae Ref. 5 - p. 529

390. THE INFERIOR GLUTEAL NERVE IS A POSTAXIAL BRANCH OF THE SACRAL PLEXUS WITH FIBERS FROM:
A. L-3, L-4 and L-5
B. L-4, L-5 and S-1
C. L-5, S-1 and S-2
D. S-1, S-2 and S-3
E. S-2, S-3 and S-4 Ref. 5 - p. 531

391. THE LONGEST MUSCLE IN THE BODY IS THE:
A. Quadratus femoris
B. Gastrocnemius
C. Sartorius
D. Obturator internus
E. Obturator externus Ref. 5 - p. 532

392. ALL OF THE FOLLOWING MUSCLES ARE INNERVATED BY THE OBTURATOR NERVE, EXCEPT FOR A VARIATION SHOWN BY THE:
A. Gracilis
B. Pectineus
C. Obturator externus
D. Adductor longus
E. Adductor magnus Ref. 5 - p. 535

393. WHICH ONE OF THE FOLLOWING IS NOT A BRANCH OF THE FEMORAL ARTERY:
A. Superficial epigastric
B. Descending genicular
C. Deep external pudendal
D. Sural
E. Superficial circumflex iliac Ref. 5 - p. 544

394. THE FEMORAL NERVE IS FORMED BY THE POSTERIOR BRANCHES OF:
A. L-1, L-2 and L-3
B. L-2, L-3 and L-4
C. L-3, L-4 and L-5
D. L-4, L-5 and S-1
E. S-1, S-2 and S-3 Ref. 5 - p. 549

SECTION III - THE LOWER EXTREMITY

395. AN INJURY TO THE UPPER LATERAL MARGIN OF THE POPLITEAL FOSSA IS MOST APT TO INJURE THE ____________ NERVE:
A. Common peroneal
B. Tibial
C. Sciatic
D. Obturator
E. Femoral Ref. 5 - p. 551

396. ALL OF THE FOLLOWING ARE MUSCLES OF THE ANTERIOR LEG COMPARTMENT, EXCEPT FOR THE:
A. Tibialis anterior
B. Extensor hallucis longus
C. Extensor digitorum longus
D. Peroneus tertius
E. Peroneus brevis Ref. 5 - p. 553

397. THE GREATER BULK OF THE CALF IS FORMED BY THE:
A. Peroneus longus
B. Popliteus
C. Gastrocnemius
D. Tibialis posterior
E. Flexor digitorum longus Ref. 5 - p. 556

398. THE MUSCLES OF THE SUPERFICIAL POSTERIOR COMPARTMENT OF THE LEG ALL INSERT INTO THE:
A. Tibia
B. Fibula
C. Talus
D. Calcaneus
E. Lunate Ref. 5 - p. 559

399. THE TIBIA IS OSSIFIED FROM ____________ CENTERS:
A. 1
B. 2
C. 3
D. 4
E. 5 Ref. 5 - p. 561

400. IN THE MALE, FUSION OF THE EPIPHYSES OF THE LOWER EXTREMITY TAKES PLACE AT APPROXIMATELY:
A. 6 months
B. 4 years
C. 8 years and 6 months
D. 10 years and 6 months
E. 16 years and 6 months Ref. 5 - p. 562

401. THE DEEP FASCIA ENVELOPING THE THIGH IS CALLED:
A. Fascia lata
B. Darto's fascia
C. Fascia lumborum
D. Camper's fascia
E. Scarpa's fascia Ref. 2 - p. 360

SECTION III - THE LOWER EXTREMITY

402. THE SAPHENOUS NERVE COMES TO THE SURFACE BETWEEN THE:
A. Obturator internus and obturator externus
B. Gluteus medius and gluteus externus
C. Gluteus minimus and sartorius
D. Sartorius and gracilis
E. Piriformis and sartorius Ref. 2 - p. 362

403. THE BASE OF THE FEMORAL TRIANGLE (OF SCARPA) IS FORMED BY THE:
A. Sartorius
B. Adductor longus
C. Inguinal ligament
D. Pubic tubercle
E. None of the above Ref. 2 - p. 366

404. THE FEMORAL VEIN LIES ____________ TO THE FEMORAL ARTERY IN THE FEMORAL SHEATH:
A. Lateral
B. Medial
C. Anterior
D. Posterior
E. Femoral vein is not in the femoral sheath
Ref. 2 - p. 367

405. THE RECTUS FEMORIS MUSCLE ASSISTS IN ____________ THE KNEE:
A. Adducting
B. Abducting
C. Flexing
D. Extending
E. Rotating Ref. 2 - p. 371

406. WHEN SITTING, THE WEIGHT OF THE BODY IS SUPPORTED BY THE:
A. Pubic arch
B. Ischial tuberosities
C. Ramus of the ischium
D. Body of the ischium
E. Body of the pubis Ref. 2 - p. 375

407. THE OBTURATOR NERVE IS DERIVED FROM THE ANTERIOR RAMI OF:
A. T-12, L-1 and L-2
B. L-1, L-2 and L-3
C. L-2, L-3 and L-4
D. L-3, L-4 and S-1
E. None of the above Ref. 2 - p. 392

408. THE CHIEF FUNCTION OF THE MEDIAL CIRCUMFLEX ARTERY IS TO SUPPLY THE:
A. Gluteus maximus
B. Gluteus minimus
C. Hamstring muscles
D. Head and neck of the femur
E. None of the above Ref. 2 - p. 393

SECTION III - THE LOWER EXTREMITY

409. THE TIBIAL COLLATERAL LIGAMENT (MEDIAL LIGAMENT OF THE KNEE) EXTENDS FROM THE MEDIAL EPICONDYLE OF THE FEMUR TO THE:
A. Lateral condyle of the tibia
B. Medial condyle of the tibia
C. Articular capsule
D. Neck of the fibula
E. Neck of the tibia Ref. 2 - p. 398

410. THE LOWER END OF THE TIBIA IS EXPANDED AND OFFERS A BEARING SURFACE FOR THE:
A. Calcaneus
B. Femur
C. Pisiform
D. Talus
E. Fibula Ref. 2 - p. 399

411. TO EXPOSE THE ANTERIOR TIBIAL RECURRENT ARTERY, THE ___________ MUSCLE MUST BE SEVERED:
A. Tibialis anterior
B. Extensor digitorum brevis
C. Extensor hallicus brevis
D. Extensor digitorum longus
E. Gluteus maximus Ref. 2 - p. 400

412. THE DEEP PERONEAL NERVE ACCOMPANIES THE ___________ ARTERY THROUGH THE LEG:
A. Arcuate
B. Dorsal digital
C. Femoral
D. Saphenous
E. Anterior tibial Ref. 2 - p. 401

413. THE MOST POWERFUL EXTENSOR OF THE THIGH IS THE:
A. Gluteus maximus
B. Psoas major
C. Iliacus
D. Obturator externus
E. Piriformis Ref. 4 - p. 569

414. WHICH ONE OF THE FOLLOWING IS A "RIBBON-LIKE" MUSCLE?:
A. Adductor longus
B. Gluteus maximus
C. Psoas major
D. Piriformis
E. Sartorius Ref. 4 - p. 575

415. THE QUADRICEPS FEMORIS IS COMPOSED OF ALL OF THE FOLLOWING, EXCEPT THE:
A. Rectus femoris
B. Gracilis
C. Vastus lateralis
D. Vastus intermedius
E. Vastus medialis Ref. 4 - p. 576

SECTION III - THE LOWER EXTREMITY

416. WHICH OF THE FOLLOWING IS NOT AN ACTION OF THE GRACILIS?:
A. Adduct the thigh
B. Flex the leg
C. Extend the thigh
D. Laterally rotate the thigh
E. All of these Ref. 4 - p. 580

417. THE POPLITEAL SPACE IS BOUNDED BELOW BY THE:
A. Gastrocnemius
B. Semitendinosus
C. Semimembranosus
D. Biceps
E. Sartorius Ref. 4 - p. 584

418. ALL OF THE FOLLOWING MUSCLES DORSIFLEX THE FOOT EXCEPT FOR THE:
A. Quadriceps femoris
B. Tibialis anterior
C. Extensor digitorum longus
D. Peroneus tertius
E. Extensor hallucis longus Ref. 4 - p. 587

419. THE EXTERNAL PUDENDAL ARTERY ARISES FROM THE:
A. Superficial epigastric
B. Superficial circumflex iliac
C. Inguinal
D. Iliolumbar
E. Femoral Ref. 4 - p. 769

420. THE PERONEAL ARTERY ARISES FROM THE:
A. Anterior tibial
B. Posterior tibial
C. Popliteal
D. Femoral
E. Tibial nutrient Ref. 4 - p. 778

421. THE LATERAL ASPECT OF THE THIGH HAS A CUTANEOUS NERVE SUPPLY DERIVED FROM:
A. L-2, L-3
B. L-3, L-4
C. L-5, S-1, S-2
D. L-4, L-5, S-1
E. S-1, S-2 Ref. 4 - pp. 1092-1093

422. THE SOLE OF THE FOOT HAS A CUTANEOUS NERVE SUPPLY DERIVED FROM:
A. L-4
B. L-5
C. S-1
D. S-2
E. All of these Ref. 4 - p. 1093

SECTION III - THE LOWER EXTREMITY

423. THE DESCENDING GENICULAR ARTERY IS THE LAST BRANCH OF THE:
A. Popliteal
B. Posterior tibial
C. Femoral
D. Superficial epigastric
E. Anterior tibial Ref. 4 - p. 773

424. THE MOST SUPERFICIAL STRUCTURE FOUND WITHIN THE POPLITEAL FOSSA IS THE:
A. Popliteal vein
B. Popliteal artery
C. Profunda artery
D. Tibial nerve
E. Posterior tibial artery Ref. 4 - p. 774

425. THE SURAL ARTERIES ARISE FROM THE:
A. Popliteal
B. Anterior tibial
C. Posterior tibial
D. Femoral
E. Profunda Ref. 4 - p. 774

426. THE ADDUCTOR CANAL BEGINS AT THE APEX OF THE:
A. Popliteal fossa
B. Inguinal canal
C. Femoral triangle
D. Sciatic notch
E. Hiatus saphenous Ref. 3 - p. 103

427. THE ONLY NERVE OF THE LOWER EXTREMITY WHICH CAN BE LOCATED BY PALPATION IS THE ____________ AS IT WINDS AROUND THE LATERAL SIDE OF THE NECK OF THE FIBULA:
A. Femoral
B. Sciatic
C. Common peroneal
D. Anterior tibial
E. Saphenous Ref. 3 - p. 105

428. ATTACHED TO THE FOVEA CAPITIS FEMORIS IS THE:
A. Ligamentum acetabulare
B. Galea aponeurotica
C. Linea aspera
D. Pectineal ligament
E. Sacrotuberal ligament Ref. 3 - p. 247

429. A SLIGHT RIDGE, THE LINEA QUADRATA, OF THE FEMUR RECEIVES THE INSERTION OF:
A. Piriformis
B. Psoas major
C. Gluteus medius
D. All of the above
E. None of the above Ref. 3 - p. 248

SECTION III - THE LOWER EXTREMITY

430. THE FEMUR IS OSSIFIED FROM ____________ CENTERS:
A. 1
B. 2
C. 3
D. 4
E. 5 Ref. 3 - p. 254

431. THE PATELLA ARTICULATES WITH THE:
A. Femur
B. Tibia
C. Fibula
D. Femur and fibula
E. Tibia and fibula Ref. 3 - p. 256

432. THERE ARE ____________ TARSAL BONES, THE LARGEST OF WHICH IS THE:
A. 5, talus
B. 7, calcaneus
C. 7, cuboid
D. 5, navicular
E. 7, talus Ref. 3 - pp. 264, 266

433. THE TALUS ARTICULATES WITH ALL OF THE FOLLOWING, EXCEPT THE:
A. Tibia
B. Fibula
C. Calcaneus
D. Navicular
E. Cuboid Ref. 3 - p. 266

434. THE CALCANEUS ARTICULATES WITH THE:
A. Tibia
B. Fibula
C. Cuboid
D. Navicular
E. Cuneiform Ref. 3 - p. 268

435. WHICH ONE OF THE FOLLOWING IS NOT CHARACTERISTIC OF THE FIRST METATARSAL BONE:
A. Very thick
B. Longest of the metatarsal bones
C. Body is strong
D. Head is large
E. Medially receives insertion of part of the tendon of the tibialis anterior Ref. 3 - p. 271

SECTION III - THE LOWER EXTREMITY

ANSWER THE FOLLOWING QUESTIONS ACCORDING TO THE KEY BELOW:
A. If both statement and reason are true and related cause and effect
B. If both statement and reason are true but not related cause and effect
C. If the statement is true but the reason is false
D. If the statement is false but the reason is true
E. If both statement and reason are false

436. The femur is the largest and longest bone of the skeleton
because
It transmits the weight of the trunk from the hip bone to the tibia.
Ref. 4 - p. 282

437. The patella is a sesamoid bone
because
Ossification is completed at the time of birth.
Ref. 4 - pp. 287-288

438. In compressing the femoral artery to secure hemostasis care must be taken not to compress the femoral vein
because
The femoral vein lies lateral to the artery in the femoral sheath.
Ref. 4 - p. 767

439. The popliteal artery is a continuation of the femoral artery
because
It runs through the popliteal fossa. Ref. 4 - p. 773

440. Unless absolutely necessary, ligation of the posterior tibial artery should not be attempted
because
Gangrene occurs in a large percentage of cases following such a ligation.
Ref. 4 - pp. 775-776

441. The dorsalis pedis artery is a continuation of the posterior tibial artery
because
It has no branches in the foot. Ref. 4 - p. 777

442. The great saphenous vein passes behind the lateral epicondyle
because
It is a superficial vein of the lower extremity.
Ref. 4 - pp. 823-825

443. The subcutaneous capillary (lymphatic) plexus of the lower extremity is most abundant in the soles of the feet
because
An analogous situation exists in the upper extremity.
Ref. 4 - p. 902

444. Infections of the knee joint may be manifested by enlarged subinguinal lymph nodes
because
Some superficial vessels follow the great saphenous vein to end in these nodes. Ref. 4 - p. 905

SECTION III - THE LOWER EXTREMITY

445. All the proper digital nerves run along the dorsum of the toes
because
They lie dorsal to the corresponding arteries.
Ref. 4 - p. 1084

446. The common peroneal nerve descends along the posterior border of the biceps femoris muscle
because
It leaves the popliteal space at the lateral angle.
Ref. 4 - p. 1085

447. The homology of the sacral and brachial plexuses is not well carried out in distribution of nerves
because
The radial nerve supplies the palm and palmar muscles but the greater sciatic nerve does not supply the sole of the foot.
Ref. 4 - pp. 1059-1060, 1080-1083

448. In embryos, the development of the leg precedes that of the arm
because
In postnatal life the upper limb increases in length more rapidly than the lower. Ref. 4 - p. 78

449. The hip bone can be considered as composed of only two parts: the ilium and ischium
because
The pubis does not help form the acetabulum.
Ref. 4 - p. 273

450. The greater sciatic notch of the ischium is converted into a foramen
because
It transmits the femoral nerve and artery. Ref. 4 - p. 277

451. The sciatic tuber supports the body in the sitting position
because
In the erect position the tuber is covered by the inferior border of the gluteus maximus muscle. Ref. 4 - p. 277

452. The tibia alone transmits the weight of the trunk to the foot
because
It is the largest bone in the skeleton. Ref. 4 - p. 289

453. In man the fibula can be regarded as a rudimentary bone
because
It participates in the formation of the ankle joint but not the knee joint.
Ref. 4 - pp. 294-295

454. In fractures of the leg, the fibula usually snaps at a more proximal level than the tibia
because
The fibula bends more than does the tibia. Ref. 4 - pp. 294-295

SECTION III - THE LOWER EXTREMITY

455. The talus is ossified from four centers
because
It articulates with four bones. Ref. 4 - pp. 295-297

456. The first metatarsal is the most modified of all the metatarsal bones
because
It is the shortest, thickest, strongest and most massive.
Ref. 4 - p. 304

457. The glenoid lip (of fibrocartilage) gives the acetabulum a depth exceeding a hemisphere
because
The hip joint permits mixed gliding, rolling and rotation.
Ref. 4 - pp. 388-390

458. The medial meniscus of the knee joint is more frequently displaced
because
Of its greater fixity, its greater reactions to strain and weaker attachments to the transverse ligaments. Ref. 4 - p. 395

459. With infections of the thigh, pus usually does not break through to the surface
because
The fascia lata is very dense. Ref. 4 - pp. 563, 567, 573

460. The Gracilis, Pectineus, Adductor longus and brevis, Adductor magnus and Obturator externus all adduct the thigh
because
They are all innervated by the obturator nerve.
Ref. 4 - pp. 579-581

ANSWER THE FOLLOWING QUESTIONS ACCORDING TO THE KEY BELOW:
A. If only A is correct
B. If only B is correct
C. If A and B are correct
D. If A and B are incorrect

461. PROMINENT MUSCLES ON THE ANTERIOR SURFACE OF THE THIGH:
A. Quadriceps femoris
B. Sartorius
C. Both
D. Neither Ref. 3 - p. 99

462. THE PROMINENCE OF THE BUTTOCK IS CAUSED BY THE:
A. Gluteus maximus
B. Quadratus minimus
C. Both
D. Neither Ref. 3 - p. 99

SECTION III - THE LOWER EXTREMITY

463. THE ADDUCTOR CANAL:
A. Occupies the lower one-third of the thigh
B. Begins at the apex of the femoral triangle
C. Both
D. Neither Ref. 3 - p. 103

464. IN THE POPLITEAL FOSSA ARE THE:
A. Popliteal artery
B. Popliteal nerve
C. Both
D. Neither Ref. 3 - p. 103

465. INSERTING INTO THE SUMMIT OF THE LESSER TROCHANTER OF THE FEMUR:
A. Quadratus femoris
B. Sartorius
C. Both
D. Neither Ref. 3 - p. 248

466. MUSCLES ORIGINATING FROM THE PATELLA:
A. Psoas anterior
B. Anterior tibialis
C. Both
D. Neither Ref. 3 - p. 255

467. THE TALUS:
A. Is the largest of the tarsal bones
B. Supports the tibia
C. Both
D. Neither Ref. 3 - p. 264

468. THE CALCANEUS:
A. Forms the heel
B. Provides a strong lever for the calf muscles
C. Both
D. Neither Ref. 3 - p. 266

469. THE PATELLAR LIGAMENT (ANTERIOR LIGAMENT):
A. Is part of the common tendon of the quadratus femoris
B. Extends from the patella to the tibia
C. Both
D. Neither Ref. 3 - p. 353

470. MOVEMENTS AT THE KNEE JOINT ARE:
A. Circumduction
B. Supination
C. Both
D. Neither Ref. 3 - p. 359

471. THE HIATUS SAPHENOUS (FOSSA OVALIS):
A. Is an oval aperture in fascia lata in the proximal thigh
B. Ends in the inguinal ligament
C. Both
D. Neither Ref. 3 - pp. 494-495

SECTION III - THE LOWER EXTREMITY

472. THE GRACILIS MUSCLE:
A. Abducts the thigh
B. Is innervated by the sciatic nerve
C. Both
D. Neither
Ref. 3 - p. 498

473. THE TENDON OF ACHILLES:
A. Is the common tendon of the Gastrocnemius and Soleus
B. Inserts into the talus
C. Both
D. Neither
Ref. 3 - p. 510

474. THE POPLITEUS MUSCLE:
A. Flexes the leg and rotates it medially
B. Is innervated by a branch of the tibial nerve
C. Both
D. Neither
Ref. 3 - p. 511

475. BRANCHES OF THE FEMORAL ARTERY:
A. Superficial epigastric
B. Descending genicular
C. Both
D. Neither
Ref. 3 - p. 661

476. CONCERNING THE HIP JOINT:
A. Form: diarthrosis, simple, spheroidal, enarthrosis
B. Movements permitted: mixed gliding, rolling, rotation
C. Both
D. Neither
Ref. 4 - p. 385

477. CONCERNING THE KNEE JOINT:
A. Has the same movements permitted as the hip joint
B. Is the 2nd largest joint (2nd to the hip joint)
C. Both
D. Neither
Ref. 4 -p. 391

478. RELATIONSHIPS OF THE FEMORAL ARTERY INCLUDE:
A. Lateral side: femoral vein
B. Medial side: femoral nerve
C. Both
D. Neither
Ref. 4 - pp. 767

479. BOUNDARIES OF THE ADDUCTOR (HUNTER'S) CANAL INCLUDE:
A. Anteriorly: Sartorius muscle
B. Posteriorly: Adductors longus and adductor magnus muscles
C. Both
D. Neither
Ref. 4 - p. 767

480. THE ARTERIAL ANASTOMOSIS AROUND THE KNEE JOINT IS IN PART DUE TO:
A. Popliteal anastomosis with posterior tibial
B. Superior medial genicular anastomosis with anterior tibial
C. Both
D. Neither
Ref. 4 - p. 774

SECTION III - THE LOWER EXTREMITY

481. CONCERNING THE DORSALIS PEDIS ARTERY:
A. A continuation of posterior tibial
B. No branches in the foot
C. Both
D. Neither Ref. 4 - p. 783

482. SUPERFICIAL VEINS OF THE LOWER EXTREMITY:
A. Are superficial to the deep fascia
B. Have no valves
C. Both
D. Neither Ref. 4 - p. 821

483. DEEP VEINS OF THE LOWER EXTREMITY:
A. Follow the course of the arteries
B. Have valves
C. Both
D. Neither Ref. 4 - p. 821

484. THE SUPERFICIAL INGUINAL LYMPH NODES RECEIVE THE SUBCUTANEOUS DRAINAGE OF THE:
A. External genitalia
B. Gluteal regions
C. Both
D. Neither Ref. 4 - p. 902

485. THE CUTANEOUS NERVE DISTRIBUTION FROM THE SOLE OF THE FOOT IS IN PART DERIVED FROM:
A. L-2 and L-3
B. L-4 and L-5
C. Both
D. Neither Ref. 4 - pp. 1098-1099

FOLLOWING THE CASE HISTORY ARE SEVERAL QUESTIONS BASED ON THE HISTORY. CHOOSE THE BEST ANSWER FOR EACH QUESTION:

A 26 year-old factory worker sustains a deep laceration of the middle of the right leg by a high-speed cutting machine. Emergency first aid consists of placing a tourniquet around the right thigh and applying pressure to the bleeding vessels. The patient is transported by ambulance to a nearby hospital but loses a great deal of blood while en route. In the Emergency Room he is given two units of blood and the posterior tibial artery is ligated. Examination also reveals that the tibial nerve is completely severed. The patient survives.

486. TO SUCCESSFULLY LIGATE THE ARTERY (BELOW THE ORIGIN OF THE PERONEAL), THE MUSCLES SHOULD BE RELAXED BY:
A. Extension of the knee and plantar extension of the foot
B. Flexion of knee and plantar extension of foot
C. Extension of knee and plantar flexion of foot
D. Flexion of knee and plantar flexion of foot
E. None of the above Ref. 4 - pp. 776-778

SECTION III - THE LOWER EXTREMITY

487. STRUCTURE(S) ASSOCIATED WITH THE AREA UNDER DISCUSSION:
A. Great saphenous vein
B. Soleus muscle
C. Sural vessels
D. Tibial nerve
E. All are involved Ref. 4 - pp. 776-778

488. RESULT OF LIGATION OF THE POSTERIOR TIBIAL ARTERY:
A. Never results in gangrene
B. Always results in gangrene
C. Rarely results in gangrene
D. Gangrene occurs in a small percentage of cases
E. Gangrene occurs in a large percentage of cases
Ref. 4 - pp. 776-778

489. SECTION OF THE TIBIAL NERVE AS DESCRIBED WOULD RESULT IN A LOSS OF INNERVATION OF:
A. Tibialis posterior
B. Soleus
C. Flexor digitorum longus
D. None of these muscles
E. All of these muscles
F. All of these muscles and others Ref. 4 - pp. 1082-1084

490. CUTANEOUS LOSSES WOULD BE LOCALIZED TO:
A. Medial side and plantar surface of heel
B. Lateral side and plantar surface of heel
C. Only the sole of the foot
D. Only the dorsum of the big toe
E. No cutaneous losses would occur Ref. 4 - p. 1083

SECTION III - THE LOWER EXTREMITY

IDENTIFY THE MUSCLES ORIGINATING AND INSERTING ON THE FEMUR AND PELVIC GIRDLE AS INDICATED BY THE GUIDE LINES.

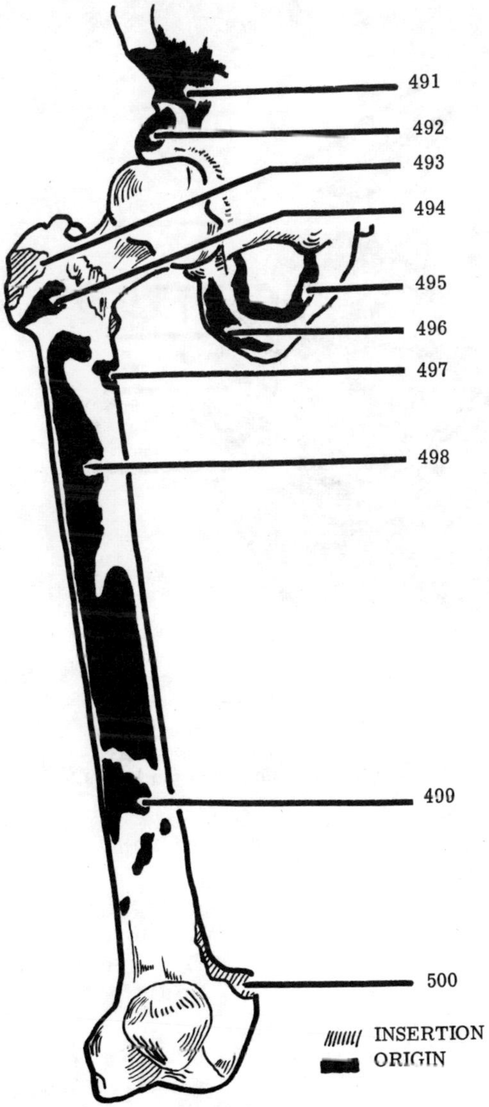

A. Rectus femoris
B. Vastus lateralis
C. Pectineus
D. Gracilis
E. Adductor magnus
F. Abductor magnus
G. Vastus intermedius
H. Ilio-psoas
I. Articular muscle
J. Adductor brevis
K. Sartorius
L. Gluteus maximus
M. Gluteus minimus
N. Vastus medialis
O. Obturator externus

Ref. 1 - p. 265

SECTION III - THE LOWER EXTREMITY

IDENTIFY THE STRUCTURES INDICATED BY THE GUIDE LINES IN THE SCHEMATIC CROSS SECTION THROUGH THE LOWER PART OF THE MIDDLE-THIRD OF THE LEG.

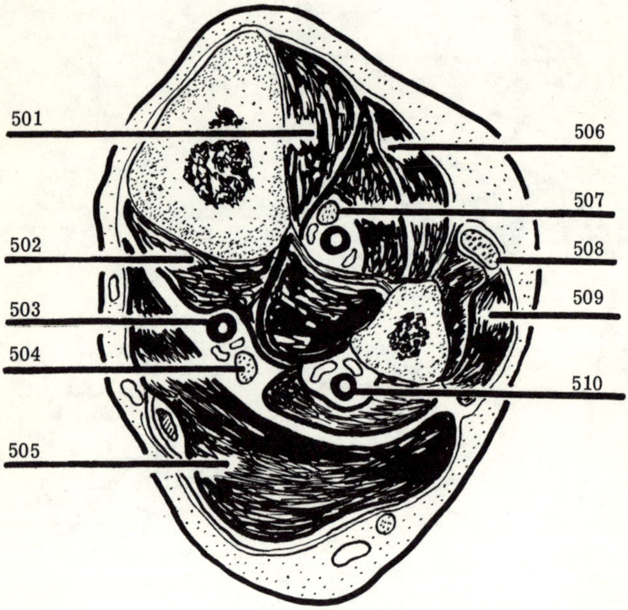

A. Gastrocnemius muscle
B. Posterior tibial artery
C. Tibialis anterior muscle
D. Peroneus brevis muscle
E. Peroneus longus muscle
F. Posterior tibial nerve
G. Femoral artery
H. Sciatic nerve
I. Tibialis posterior muscle
J. Musculo-cutaneous nerve
K. Extensor digitorum longus muscle
L. Soleus muscle
M. Flexor digitorum longus muscle
N. Deep peroneal nerve
O. Peroneal artery

Ref. 1 - p. 310

SECTION III - THE LOWER EXTREMITY

IDENTIFY THE STRUCTURES INDICATED BY THE GUIDE LINES IN THE ILLUSTRATION OF THE BOUNDARIES AND CONTENTS OF THE RIGHT FEMORAL TRIANGLE.

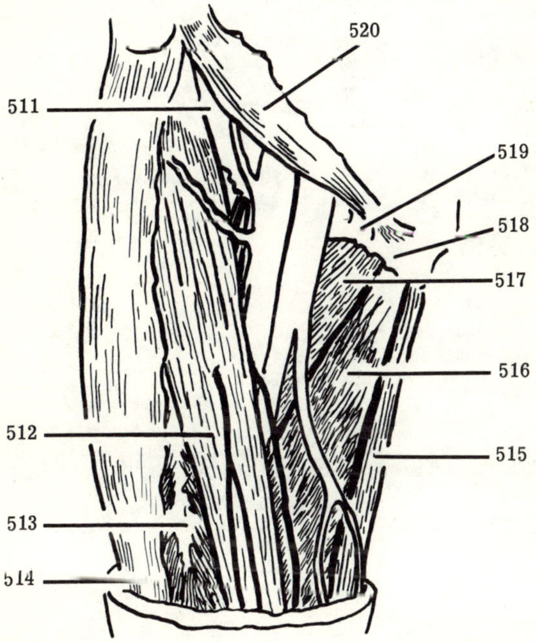

A. Iliac fascia
B. Ilio-tibial tract
C. Gracilis
D. Sartorius
E. Fascia lateralis
F. Rectus femoris
G. Vastus medialis
H. Popliteus
I. Lacunar ligament
J. Pectineus
K. Pectineal fascia
L. Iliacus
M. Adductor brevis
N. Adductor longus
O. Inguinal ligament

Ref. 1 - p. 258

SECTION IV - THE THORAX

FOR EACH OF THE FOLLOWING MULTIPLE CHOICE QUESTIONS SELECT THE APPROPRIATE ANSWER OR ANSWERS. ONE OR MORE, OR ALL OF THE CHOICES MAY BE USED IN CORRECTLY ANSWERING THE QUESTION.

521. THE FIRST RIB:
A. Is the most curved of all the ribs
B. Is usually the shortest of all the ribs
C. Is broad and flat
D. Articulates with the second thoracic vertebra
E. Provides a site of attachment for the scalenus anterior
Ref. 3 - p. 143

522. THE EXTERNAL INTERCOSTAL MUSCLES:
A. Are 12 pair in number
B. Extend from rib tubercle dorsally to cartilage of rib ventrally
C. Are thicker than the internal intercostals
D. Draw adjacent ribs apart
E. Are innervated by intercostal nerves Ref. 3 - p. 414

523. THE LEVATORES COSTARUM:
A. Raise the ribs
B. Increase the thoracic cavity
C. Extend the vertebral column
D. Bend the vertebral column laterally
E. Rotate the vertebral column slightly to the opposite side
Ref. 3 - p. 415

524. PASSING THROUGH THE AORTIC HIATUS OF THE DIAPHRAGM:
A. Aorta
B. Esophagus
C. Vagus nerve
D. Phrenic nerve
E. Thoracic duct Ref. 3 - p. 417

525. PASSING THROUGH THE ESOPHAGEAL HIATUS OF THE DIAPHRAGM:
A. Right phrenic nerve
B. Azygos artery
C. Azygos vein
D. Inferior vena cava
E. Vagus nerve Ref. 3 - p. 418

526. THE LIGAMENT OF THE LEFT VENA CAVA (VESTIGIAL FOLD OF MARSHALL):
A. Stretches from left pulmonary artery to atrial wall
B. Transmits the left coronary artery
C. Causes left ventricular hypertrophy
D. Quickens the pace of the bundle of His
E. Is found only in males Ref. 3 - p. 546

SECTION IV - THE THORAX

527. THE ASCENDING AORTA:
A. Is about 5cm in length
B. Is covered by visceral pericardium
C. Begins at the semilunar valve
D. Becomes the arch of the aorta
E. Has no branches Ref. 3 - pp. 575-576

528. THE PULMONARY ARTERY:
A. Carries arterial blood
B. Arises from the conus arteriosus of the right ventricle
C. Divides at about the level of the second left chondrosternal articulation
D. Is separated from the thymus by pericardium
E. Is the longest artery of the body Ref. 4 - pp. 663-665

529. BRANCHES OF THE AORTIC ARCH:
A. Innominate
B. Right common carotid
C. Right subclavian
D. Coronary
E. Left subclavian Ref. 4 - p. 669

530. NERVES TO THE PULMONARY PLEURA COME FROM THE:
A. Intercostals
B. Vagus
C. Phrenic
D. Sympathetic
E. Spinal accessory Ref. 4 - p. 1438

531. CONTENTS OF THE MIDDLE MEDIASTINUM INCLUDE THE:
A. Ascending aorta
B. Trunk of pulmonary artery
C. Phrenic nerves
D. Arch of azygos vein
E. Bronchial lymph nodes Ref. 4 - p. 1438

532. FOUND AT THE ROOT OF THE LUNG:
A. Bronchus
B. Pulmonary artery
C. Pulmonary vein
D. Pulmonary nerve
E. Bronchial vessels Ref. 4 - pp. 1441-1442

533. THE VEINS OF THE THYMUS DRAIN INTO THE:
A. Right innominate
B. Internal mammary
C. Superior thyroid
D. Right subclavian
E. Inferior vena cava Ref. 4 - p. 1555

SECTION IV - THE THORAX

534. THE SUPERFICIAL LYMPHATICS OF THE THORAX PASS TO THE:
A. Axillary nodes
B. Inguinal nodes
C. Internal mammary vein
D. Thoracic duct
E. Supraclavicular nodes — Ref. 4 - p. 883

535. THE POSTERIOR MEDIASTINAL NODES RECEIVE VESSELS FROM THE:
A. Esophagus
B. Larynx
C. Heart
D. Lung
E. Diaphragm — Ref. 4 - pp. 884-885

536. THE 12TH RIB HAS A:
A. Large head
B. Single facet for articulation
C. Narrow shaft
D. Costal groove
E. Neck — Ref. 4 - p. 238

537. CONCERNING THE STERNOCHONDRAL ARTICULATIONS:
A. Surfaces of costal notches coated with hyaline cartilage
B. All joints are a diarthrosis
C. Only motions are up and down
D. First 3 ribs do not articulate with sternum
E. Arterial supply from internal mammary
Ref. 4 - pp. 347-348

538. DURING THE PROCESS OF NORMAL RESPIRATIONS:
A. 90 joints move
B. Intercostal muscles do not act
C. Diaphragm relaxes during expiration
D. Expiration produced solely by internal intercostals
E. Muscles of neck aid in forced inspiration
Ref. 4 - pp. 526-529

539. INTRINSIC THORACOABDOMINAL MUSCLES:
A. Obliquus externus abdominis
B. Intercostales externi
C. Latissimus dorsi
D. Serratus posterior superior
E. Splenius capitis — Ref. 4 - p. 528

540. THE HEART RECEIVES NERVES FROM THE:
A. Parasympathetic system via the phrenics
B. Parasympathetic system via the vagi
C. Sympathetic system via the vagi
D. Sympathetic system via the phrenics
E. None of the above — Ref. 4 - pp. 653-654

SECTION IV - THE THORAX

541. RELATIONS OF THE HEART AND GREAT VESSELS:
A. Base formed by the ventricles
B. Apex in the 5th intercostal space, left side
C. Pulmonary ostium under 2nd left interspace
D. Aortic ostium largely lies substernally
E. Of the ostia, pulmonary is nearest anterior thoracic wall
Ref. 4 - p. 659

542. BRANCHES OF THE BRACHIOCEPHALIC TRUNK:
A. Thyroid ima
B. Left common carotid
C. Right common carotid
D. Left subclavian
E. Right subclavian Ref. 4 - p. 670

543. THE THORACIC DUCT:
A. Extends from 2nd lumbar vertebra to right internal jugular vein
B. Receives all lymphatics below diaphragm
C. At its origin is a dilated portion; cisterna chyli
D. All of the above
E. None of the above Ref. 4 - p. 870

544. RELATIONS OF CERVICAL AND THORACIC NERVES TO BRANCHES OF BRACHIAL PLEXUS:
A. C-7 and C-8 to thoracodorsal
B. C-8 and T-1 to medial anterior thoracic
C. C-5, C-6 and C-7 to dorsal scapular
D. C-5, C-6 and C-7 to lateral anterior thoracic
E. C-5, C-6 and C-7 to long (posterior) thoracic
Ref. 4 - p. 1059

545. THE POSTERIOR MEDIASTINUM:
A. Extends from 4th thoracic vertebra to diaphragm
B. Includes esophagus and thoracic duct
C. Includes descending thoracic aorta and azygos veins
D. Includes vagi and splanchnic nerves
E. None of the above Ref. 4 - p. 1440

546. MUSCLES ATTACHED TO THE XIPHOID PROCESS OF THE STERNUM:
A. Rectus abdominis
B. Transversus thoracis
C. Diaphragm
D. External intercostals
E. Internal intercostals Ref. 5 - p. 304

547. MUSCLES OF THE THORAX:
A. Levatores costarum
B. Serratus posterior superior
C. Serratus posterior inferior
D. Diaphragm
E. Latissimus dorsi Ref. 5 - pp. 307-308

SECTION IV - THE THORAX

548. BRANCHES OF THE INTERNAL THORACIC ARTERY:
A. Anterior mediastinal
B. Anterior intercostal
C. Internal mammary
D. Posterior axillary
E. Superior epigastric Ref. 5 - p. 312

549. THE SUPERIOR MEDIASTINUM:
A. Lies behind the trachea and esophagus
B. Is situated anterior to the upper four thoracic vertebrae
C. Is bounded laterally by the parietal pleura
D. Contains the roots of the lungs and azygos veins
E. Is bounded below by the diaphragm Ref. 5 - p. 315

550. CONTENTS OF THE MIDDLE MEDIASTINUM:
A. Arch of the aorta
B. Left common carotid artery
C. Phrenic nerves
D. Azygos veins
E. Thoracic duct Ref. 5 - p. 315

551. CONTENTS OF THE POSTERIOR MEDIASTINUM:
A. Esophagus
B. Vagus nerve
C. Thoracic duct
D. Heart
E. Sternopericardial ligament Ref. 5 - p. 315

552. THE PARIETAL PLEURA IS SUPPLIED BY TWIGS OF ARTERIES FROM:
A. The posterior intercostal
B. The internal thoracic
C. The superior phrenic
D. The anterior mediastinal
E. The superior epigastric Ref. 5 - p. 317

553. THE BLOOD SUPPLY TO THE THYMUS GLAND IS FROM THE:
A. Anterior mediastinal branches of the internal thoracic
B. Inferior thyroid
C. Posterior mediastinal branches of the azygos
D. Superior thymic branches of the phrenic
E. Inferior thymic branches of the internal mammary
Ref. 5 - p. 318

554. NERVES TO THE THYMUS GLAND COME FROM THE:
A. Phrenic
B. Mammary
C. Vagus
D. Pleuropericardial
E. Axillary Ref. 5 - p. 318

SECTION IV - THE THORAX

555. THE AORTIC ARCH:
A. Is a continuation of the ascending aorta
B. Terminates at the level of the disk between fourth and fifth thoracic vertebrae
C. Becomes the thoracic aorta
D. Lies anterior to the trachea
E. Has no branches Ref. 5 - pp. 323-324

556. A COMPARISON OF RIGHT AND LEFT BRONCHUS REVEALS THAT THE:
A. Right bronchus is longer
B. Left bronchus is straighter
C. Right bronchus is larger
D. Right bronchus is the more direct continuation of the trachea
E. Foreign bodies more apt to lodge in right bronchus
Ref. 5 - p. 341

557. GROUPS OF TRACHEOBRONCHIAL LYMPH NODES:
A. Tracheal
B. Superior tracheobronchial
C. Inferior tracheobronchial
D. Bronchopulmonary
E. Pulmonary Ref. 5 - pp. 341-342

558. THE LEFT PULMONARY ARTERY:
A. Passes horizontally anterior to descending aorta
B. Arches over left bronchus
C. Longer than the right
D. Larger than the right
E. Goes to root of left lung Ref. 5 - p. 347

559. LYMPHATICS OF THE ESOPHAGUS DRAIN REGIONALLY INTO THE:
A. Inferior deep cervical lymph nodes
B. Inferior thyroid nodes
C. Paratracheal lymph nodes
D. Posterior mediastinal nodes
E. Superior gastric nodes Ref. 5 - p. 351

560. BRANCHES OF THE THORACIC AORTA:
A. Pericardial
B. Esophageal
C. Mediastinal
D. Superior phrenic
E. Bronchial Ref. 5 - p. 352

SECTION IV - THE THORAX

ANSWER THE FOLLOWING QUESTIONS USING THE KEY AS OUTLINED BELOW:

A. If 1, 2 and 3 are correct
B. If 1 and 3 are correct
C. If 2 and 4 are correct
D. If all are correct
E. If some combination other than the above are correct
F. If all are incorrect

561. THE UPPER APERTURE OF THE THORAX IS FORMED BY:
1. 2nd thoracic vertebra
2. 2nd rib
3. 2nd costal cartilage
4. Manubrium sterni — Ref. 2 - p. 463

562. THE LOWER APERTURE OF THE THORAX IS FORMED BY:
1. Lower 6 costal cartilages
2. 12th rib
3. Xiphoid process
4. Body of 12th thoracic vertebra — Ref. 2 - pp. 463-464

563. MUSCLES ATTACHED TO THE STERNUM:
1. Diaphragm
2. Sternohyoid
3. Sternothyroid
4. Transversus thoracis — Ref. 2 - p. 467

564. AREAS WHERE PLEURAE DESCEND BELOW THE COSTAL MARGIN:
1. Right costo-vertebral angle
2. Left costo-vertebral angle
3. Right xiphi-costal angle
4. Left xiphi-costal angle — Ref. 2 - p. 477

565. AMONG THE CONTENTS OF THE MEDIASTINUM:
1. Trachea
2. Lung
3. Phrenic nerves
4. Diaphragm — Ref. 2 - p. 479

566. RELATIVE TO THE INTERNAL MAMMARY ARTERY:
1. Branch of the subclavian
2. Most readily exposed in 1st or 2nd intercostal space
3. Terminal branches; superior epigastric and musculophrenic
4. An intersegmental artery — Ref. 2 - pp. 472-473

567. CONTENTS OF THE ANTERIOR MEDIASTINUM:
1. Trachea
2. Vagus nerves
3. Phrenic nerves
4. Esophagus — Ref. 2 - p. 479

SECTION IV - THE THORAX

568. STRUCTURES COVERED WITH LEFT MEDIASTINAL PLEURA:
1. Thoracic duct
2. Aortic arch
3. Trachea
4. Esophagus Ref. 2 - p. 482

569. FEATURES OF THE THORACIC PORTION OF THE SYMPATHETIC TRUNK:
1. Lies within the mediastinum
2. Covered throughout with costal pleura
3. No thoracic splanchnic nerves
4. Eleven thoracic ganglia Ref. 2 - p. 486

570. CONCERNING THE RIGHT ATRIUM OF THE HEART:
1. Demarcated from right ventricle by coronary sulcus
2. Superior vena cava opens into this chamber
3. Fossa ovalis on medial wall
4. Has only 3 openings Ref. 2 - pp. 511-512

571. SO-CALLED "PECULIAR" RIBS ARE THE:
1. 1st because it has no head
2. 2nd because it is the broadest and most curved
3. 10th because it has no articular facet
4. 12th because it has no tubercle, angle, neck or costal groove
Ref. 4 - p. 239

572. THE STERNUM:
1. Consists of 3 parts
2. Consists of 6 segments in young subjects
3. Is inclined obliquely from above caudad and ventrad
4. Is not of equal width throughout Ref. 4 - pp. 240-242

573. MUSCLES ABDUCTING AND EXTENDING THE THORACIC VERTEBRAL COLUMN:
1. Iliocostales
2. Longissimus dorsi
3. Levatores costarum
4. Sacrospinalis Ref. 4 - p. 527

574. MUSCLES OF EXPIRATION:
1. Intercostales interni (costal parts)
2. Diaphragm
3. Transversus thoracis
4. Intercostales externi Ref. 4 - p. 527

575. CONCERNING THE HEART:
1. Sulcus coronarius externally separates atria from ventricles
2. Sulcus terminalis connects right sides of superior and inferior vena cava onto back of right atrium
3. Ostium venosum of left ventricle guarded by tricuspid valve
4. Crista terminalis corresponds to sulcus coronarius
Ref. 4 - pp. 614-621

SECTION IV - THE THORAX

576. FEATURES OF THE PULMONARY ARTERIES:
1. Carry arterial blood
2. Arise from the conus arteriosus
3. Left artery longer than the right
4. Left artery smaller caliber than the right
Ref. 4 - pp. 663-665

577. THE ASCENDING AORTA:
1. Begins at the upper part of the right ventricle
2. Is not enclosed within the pericardial cavity
3. Ascends to the upper border of the left 4th chondrosternal articulation
4. Measures about 2.5 cm. Ref. 4 - pp. 665-667

578. PRESSURE SYMPTOMS SECONDARY TO ANEURYSM OF THE AORTIC ARCH MAY INVOLVE THE:
1. Trachea
2. Left bronchus
3. Esophagus
4. Thoracic duct Ref. 4 - p. 666

579. INCLUDED AMONG THE DEEP VEINS OF THE THORAX ARE THE:
1. Pulmonary
2. Superior vena cava
3. Innominate
4. Thoraco-epigastric Ref. 4 - p. 819

580. CONCERNING THE LYMPHATICS OF THE THORACIC SURFACE OF THE DIAPHRAGM:
1. Rich anastomoses between two halves
2. Anastomosis with lymphatics of costal pleura
3. Richest network over central pericardial region
4. Arranged like spokes of a wheel, radiating out between muscle bundles
Ref. 4 - p. 887

581. LYMPHATICS OF THE LUNGS ARE ARRANGED:
1. In 2 sets: deep and superficial
2. Into 3 groups of deep lymphatics: bronchial, arterial, venous
3. In a superficial group on the lung surface
4. So that superficial and deep lymphatics communicate
Ref. 4 - pp. 887-888

582. COMPARISON OF THE MALE AND FEMALE THORAX REVEALS THAT THE:
1. Capacity of female thorax is less
2. Female sternum is longer
3. Cranial margin of sternum of female is lower
4. Fewer ribs in female Ref. 3 - p. 133

583. THE SERRATUS POSTERIOR SUPERIOR:
1. Is a thin, quadrilateral muscle
2. In part takes origin from ligamentum nuchae
3. Inserts into Ribs 2, 3, 4 and 5
4. Raises the ribs to which it is attached Ref. 3 - p. 415

SECTION IV - THE THORAX

584. PASSING THROUGH THE VENA CAVAL FORAMEN OF THE DIAPHRAGM:
1. Vagus nerve
2. Inferior vena cava
3. Thoracic duct
4. Some branches of the right phrenic nerve

Ref. 3 - p. 418

585. THE PHRENIC NERVE FROM THE CERVICAL PLEXUS CONTAINS FIBERS FROM:
1. C-2
2. C-3
3. C-4
4. C-5

Ref. 3 - p. 418

FOLLOWING EACH CASE HISTORY ARE SEVERAL QUESTIONS BASED ON THE HISTORY. CHOOSE THE BEST ANSWER FOR EACH QUESTION:

A 64 year-old retired carpenter is suddenly seized with excruciating chest pains, following climbing two flights of stairs. A physician is called who administers morphine for the pain and recommends hospitalization. The patient refuses and insists on remaining at home. Several hours later a second "attack" occurs and another physician is called. During the course of the examination, the patient has a cardiac arrest. The physician attempts closed cardiac massage, intracardiac adrenaline and mouth-to-mouth breathing. Eventually the patient is pronounced dead; an autopsy is performed.

586. THE PATIENT'S HEART WEIGHT 540 GRAMS; THE NORMAL WEIGHT OF A MALE HEART IS GENERALLY:
A. 120 to 160 grams
B. 180 to 240 grams
C. 280 to 340 grams
D. 360 to 450 grams
E. 500 to 600 grams

Ref. 3 - p. 543

587. EXAMINATION OF THE HEART REVEALS AN OLD AND A NEW AREA OF DAMAGE (INFARCTION) CAUSED BY OCCLUSION OF THE CIRCUMFLEX BRANCH OF THE LEFT CORONARY ARTERY. THE DAMAGE IS IN THE:
A. Left atrium and left ventricle
B. Right atrium and right ventricle
C. Left auricle and interventricular septum
D. Apex of the heart
E. Pulmonary artery

Ref. 3 - p. 578

588. THE LEFT CORONARY ARTERY BIFURCATES INTO THE CIRCUMFLEX BRANCH AND THE:
A. Anterior cardiac artery
B. Anterior interventricular artery
C. Marginal artery
D. Posterior interventricular artery
E. Sinus of Valsalva

Ref. 3 - p. 578

SECTION IV - THE THORAX

589. THE PERICARDIAL SAC IS FILLED WITH A STRAW-COLORED FLUID INDICATING PREVIOUS HEART DISEASE. THIS FLUID COULD HAVE BEEN REMOVED SAFELY BY INTRODUCING A NEEDLE IN AN AREA BOUNDED BY THE:
A. Left mid-clavicular line and third intercostal space
B. Right mid-axillary line and fourth or fifth intercostal space
C. Suprasternal notch
D. Left parasternal border and third intercostal space
E. Caudal portion of the body of the sternum and the medial ends of the left fourth and fifth costal cartilages Ref. 3 - p. 546

590. THE APEX OF THE HEART IS USUALLY LOCATED:
A. Deep to the left fifth intercostal space, 8 or 9 cm. from the midsternal line
B. Deep to the left third intercostal space, mid-clavicular line
C. Fourth intercostal space, posterior left mid-clavicular line
D. Fourth thoracic vertebra, left posterior axillary line
E. Deep to the right fourth intercostal space, parasternal border
Ref. 3 - p. 549

SECTION IV - THE THORAX

IDENTIFY THE STRUCTURES INDICATED BY THE GUIDE LINES IN THE SCHEMATIC TRANSVERSE SECTION THROUGH THE THORAX.

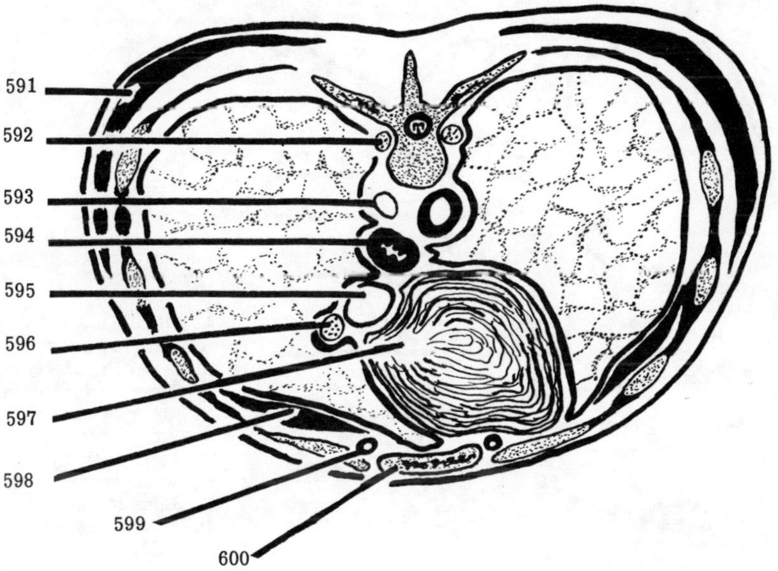

A. Pericardial sac
B. Sternum
C. Esophagus
D. Superior epigastric artery
E. Thoracic duct
F. Trachea
G. Internal mammary artery
H. Aorta
I. Left phrenic nerve
J. Latissimus dorsi muscle
K. Right phrenic nerve
L. Obliquus externus muscle
M. Inferior vena cava
N. Transversus thoracis muscle
O. Sympathetic trunk

Ref. 1 - p. 431

SECTION IV - THE THORAX

IDENTIFY THE CORONARY ARTERIES AND CARDIAC VEINS AS INDICATED BY THE GUIDE LINES.

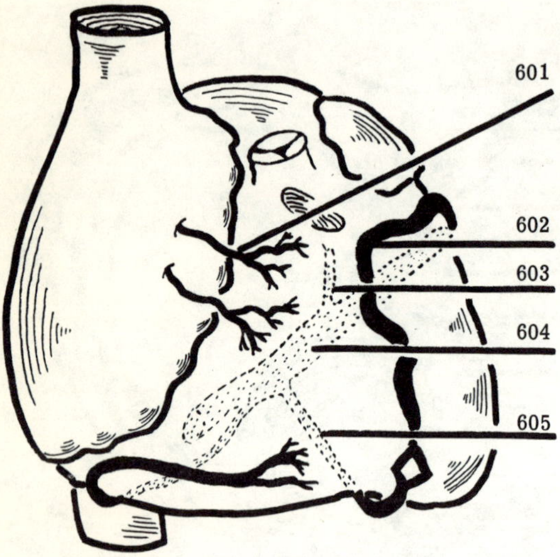

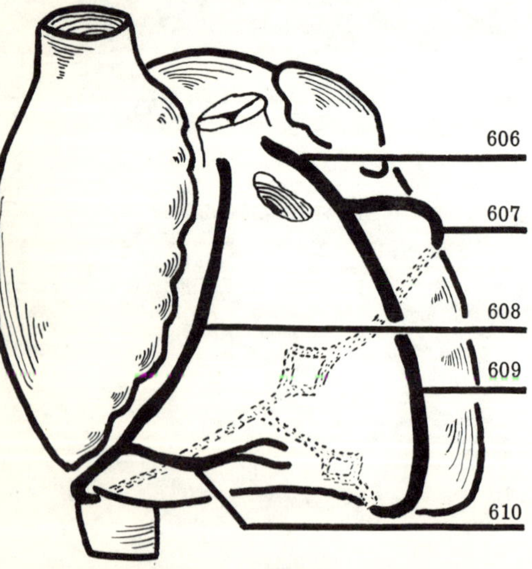

A. Middle cardiac vein
B. Marginal artery
C. Great cardiac vein
D. Superior vena cava
E. Anterior cardiac vein
F. Left coronary artery
G. Small cardiac vein
H. Sinus of Valsalva
I. Right coronary artery
J. Posterior descending artery
K. Interventricular artery
L. Inferior vena cava
M. Oblique vein
N. Circumflex artery
O. Coronary sinus

Ref. 1 - pp. 437-438

SECTION IV - THE THORAX

IDENTIFY THE MUSCLES THAT TAKE ORIGIN AND INSERT ON THE FIRST AND SECOND RIBS, AS INDICATED BY THE GUIDE LINES.

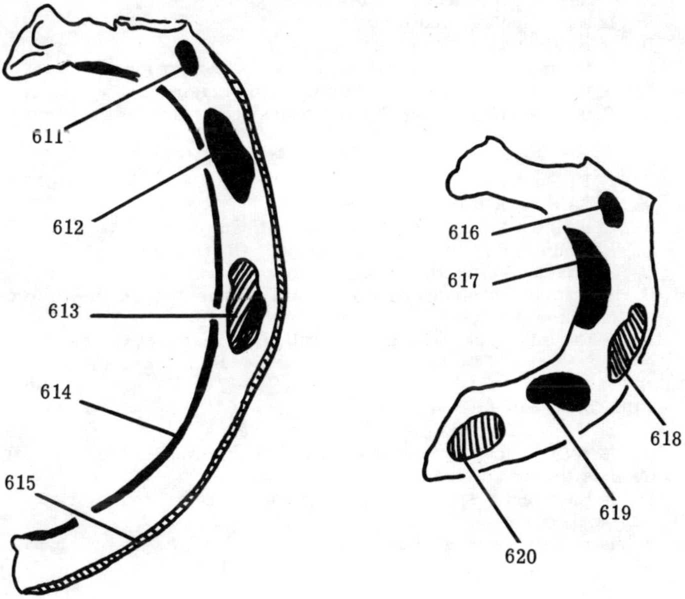

A. Serratus anterior
B. Scalenus medius
C. Subclavius
D. Rhomboideus
E. Scalenus anterior
F. Latissimus dorsi
G. Levator costae
H. Sternohyoideus
I. Thyrohyoideus
J. Internal intercostal
K. Serratus lateralis
L. Scalenus posterior
M. Pectoralis major
N. Pectoralis minor
O. External intercostal

Ref. 4 - p. 237

SECTION V - THE ABDOMEN, PERINEUM AND PELVIS

FOR EACH OF THE FOLLOWING MULTIPLE CHOICE QUESTIONS SELECT THE APPROPRIATE ANSWER OR ANSWERS. ONE OR MORE, OR ALL OF THE CHOICES MAY BE USED IN CORRECTLY ANSWERING THE QUESTION:

621. THE SUPERFICIAL LAYER OF THE SUBCUTANEOUS FASCIA OF THE ABDOMEN (CAMPER'S FASCIA):
A. Is a panniculus adiposus
B. Is continuous over the inguinal ligament
C. In the male continues down on the penis and scrotum
D. In the female is continued into the labia majora
E. Becomes continuous with the subcutaneous fascia of the perineum
Ref. 3 - p. 420

622. THE DEEP LAYER OF SUBCUTANEOUS FASCIA OF THE ABDOMEN (SCARPA'S FASCIA):
A. Contains a large quantity of adipose tissue
B. Is composed primarily of collagenous fibers
C. Is attached to the linea alba
D. Inferiorly passes over the inguinal ligament and attaches to it or to the fascia lata
E. In the thigh and perineum it is called the fascia of Colles
Ref. 3 - p. 420

623. THE LACUNAR LIGAMENT:
A. Is the medial end of the inguinal ligament
B. Is rolled under the spermatic cord and attached to the pectineal line
C. Has its apex medially at the pubic tubercle
D. In the erect position has the spermatic cord on its posterior, inferior surface
E. Is closely associated with the medial wall of the femoral canal
Ref. 3 - p. 423

624. THE PECTINEAL LIGAMENT (COOPER'S LIGAMENT):
A. Forms the posterior wall of the adductor canal
B. Continues laterally from the lacunar ligament along the pectineal line of the pubis
C. Is aponeurotic in origin
D. Is attached to the pectineal fascia
E. Is attached to the transversalis fascia Ref. 3 - p. 425

625. THE CREMASTER MUSCLE:
A. Draws the testis up toward the superficial inguinal ring
B. Is innervated by the internal pudendal nerve
C. Arises from the middle of the inguinal ligament as a continuation of the rectus abdominis muscle
D. Inserts into the testes
E. Is homologous to the Fallopian tube in the female
Ref. 3 - p. 426

SECTION V - THE ABDOMEN, PERINEUM AND PELVIS

626. THE RECTUS ABDOMINIS MUSCLE:
- A. Is separated from its counterpart on the opposite side by the linea alba
- B. Is broader but thinner inferiorly
- C. Originates along the inferior border of the twelfth rib
- D. Inserts into the rim of the pelvis
- E. Tenses the anterior abdominal wall Ref. 3 - pp. 427-428

627. ENCLOSED IN THE SHEATH OF THE RECTUS ABDOMINIS:
- A. Pyramidalis muscle
- B. Superior epigastric artery
- C. Inferior epigastric artery
- D. Superior epigastric vein
- E. Branches of intercostal nerves Ref. 3 - p. 429

628. HESSELBACH'S TRIANGLE IS BOUNDED BY THE:
- A. Inguinal ligament
- B. Genitofemoral branch of the femoral nerve
- C. Lateral border of rectus abdominis muscle
- D. Linea alba
- E. Inferior deep epigastric artery Ref. 3 - p. 432

629. PASSING THROUGH THE INGUINAL CANAL:
- A. Spermatic cord
- B. Ilioinguinal nerve
- C. Pyramidalis muscle
- D. Cremaster muscle
- E. Round ligament Ref. 3 - p. 433

630. THE PELVIC DIAPHRAGM IS COMPOSED OF MUSCLES AND THE FASCIA INVESTING THEM; THESE MUSCLES INCLUDE:
- A. Levator ani
- B. Piriformis
- C. Obturator internus
- D. Gluteus maximus
- E. Coccygeus Ref. 3 - p. 434

631. THE BLOOD SUPPLY OF THE SCROTUM IS DERIVED FROM THE:
- A. External pudendal
- B. Internal pudendal
- C. Superior epigastric
- D. Inferior epigastric
- E. All of the above Ref. 2 - p. 212

632. THE JEJUNUM DIFFERS FROM THE ILEUM IN THAT THE FORMER HAS:
- A. Less of a digestive surface
- B. A small diameter
- C. Fewer villi
- D. Larger villi
- E. A thicker wall Ref. 2 - p. 222

SECTION V - THE ABDOMEN, PERINEUM AND PELVIS

633. CONCERNING THE FALCIFORM LIGAMENT OF THE LIVER:
A. Round ligament in its free border
B. Attached to anterior abdominal wall and diaphragm
C. Prevents displacement of liver to right
D. Two openings for portal vein and hepatic artery
E. All of the above Ref. 2 - pp. 227-228

634. ORGANS FORMING THE VISCERAL SURFACE OF THE SPLEEN:
A. Stomach
B. Pancreas
C. Kidney
D. Duodenum
E. Colon Ref. 2 - p. 229

635. FIBROUS CORDS RADIATING FROM THE UMBILICUS:
A. Urachus
B. Umbilical vein
C. Umbilical arteries
D. All of the above
E. None of the above Ref. 2 - pp. 234-239

636. OBLITERATED PERITONEAL FOLDS ACTING AS MESENTERIES:
A. Lesser omentum
B. Broad ligament of uterus
C. Median umbilical ligament
D. Mesocolon
E. All of the above Ref. 2 - p. 236

637. THE PORTAL VEIN IS FORMED BY THE:
A. Renal
B. Superior mesenteric
C. Gastric
D. Splenic
E. Inferior mesenteric Ref. 2 - pp. 237-238

638. CONCERNING THE LESSER OMENTUM:
A. Extends from lesser curvature of stomach to duodenum
B. Cystic duct runs in its free edge
C. Pierced by celiac artery
D. Lying between its 2 layers are the anterior and posterior vagal trunks
E. None of the above Ref. 2 - p. 242

639. BRANCHES OF THE HEPATIC ARTERY:
A. Right gastric
B. Splenic
C. Pancreatic
D. Left gastro-epiploic
E. Gastro-duodenal Ref. 2 - p. 244

SECTION V - THE ABDOMEN, PERINEUM AND PELVIS

640. FEATURES OF THE PORTAL VEIN:
A. Drains only the gastrointestinal tract and its unpaired glands (except the liver)
B. Ascends in front of the neck of the pancreas
C. Lies behind the inferior vena cava
D. No functioning valves in portal system
E. All of the above Ref. 2 - pp. 246-247

641. PORTA-CAVAL ANASTOMOSES OCCUR BETWEEN:
A. Esophageal branches of left gastric and esophageal branches of azygos
B. Superior rectal and inferior rectal
C. Para-umbilical
D. Colic and splenic
E. All of the above Ref. 2 - p. 247

642. BOUNDARIES OF THE ISCHIO-RECTAL FOSSA INCLUDE:
A. Laterally by levator ani
B. Medially by ischium
C. Posteriorly by sacro-tuberous ligament
D. Anteriorly by base of "urogenital" diaphragm
E. All of the above Ref. 2 - p. 301

643. CONTENTS OF DEEP PERINEAL POUCH IN THE MALE:
A. Sphincter urethrae
B. Membranous urethra
C. Bulbo-urethral glands
D. All of the above
E. None of the above Ref. 2 - p. 305

644. OPENING INTO THE VESTIBULE OF THE VAGINA:
A. Urethra
B. Para-urethral glands
C. Greater vestibular glands (Bartholin's glands)
D. Cowper's glands
E. Ejaculatory ducts Ref. 2 - p. 309

645. VISCERAL BRANCHES OF THE INTERNAL ILIAC ARTERY:
A. Umbilical
B. Inferior vesical
C. Superior vesical
D. Middle rectal
E. Inferior gluteal Ref. 2 - p. 332

646. THE DUCTUS DEFERENS CROSSES THE:
A. Obturator nerve
B. Urethra
C. External iliac artery
D. External iliac vein
E. Ureter Ref. 2 - p. 328

SECTION V - THE ABDOMEN, PERINEUM AND PELVIS

647. THE ARTERIAL SUPPLY TO THE RECTUM IS IN PART DERIVED FROM:
A. Median sacral
B. Inferior gluteal
C. Internal pudendal
D. Internal obturator
E. Lateral coccygeal
Ref. 2 - p. 330

648. THE SACRAL PLEXUS IS DERIVED FROM:
A. L-1
B. L-2
C. L-3
D. L-4
E. L-5
Ref. 2 - pp. 333-334

649. CONCERNING THE SYMPHYSIS PUBIS JOINT:
A. Form : synarthrosis, symphysis
B. Shorter and broader in female
C. Arterial supply : in part from internal pudendal
D. Movements : rotation, circumduction, abduction and adduction
E. With pregnancy : joint may become unnaturally loose
Ref. 4 - pp. 382-383

650. THE COCCYGEUS MUSCLE IS:
A. In large part aponeurotic
B. Active in flexing the coccyx
C. Innervated by S-1 and S-2
D. Absent in 50% of cadaver dissections
E. Partly covered by sacrospinous ligament
Ref. 4 - pp. 558-559

651. ANATOMICAL FEATURES OF THE SPHINCTER ANI EXTERNUS MUSCLE:
A. Involuntary muscle
B. Innervated by S-2 and S-3
C. On contraction opens the anus
D. Inserts into skin in front of anus
E. Absent in 50% of cadaver dissections
Ref. 4 - pp. 559-560

652. THE BULBOSPONGIOSUS MUSCLE OF THE MALE:
A. Ensheaths the bulb of the penis
B. Arises from dense tissue covering the urogenital diaphragm
C. Inserts into symphysis pubis
D. Is innervated by perineal division of pudendal nerve
E. Lies within the superficial perineal pouch
Ref. 4 - p. 561

653. THE BULBOSPONGIOSUS MUSCLE OF THE FEMALE:
A. Is a sphincter of the vagina
B. Arises in part from fibrous tissue dorsal to clitoris
C. Is innervated by perineal division of pudendal nerve
D. Compresses the orifice of the vagina
E. Is deep to the skin and superficial fascia
Ref. 4 - p. 561

SECTION V - THE ABDOMEN, PERINEUM AND PELVIS

654. THE DUODENUM IS:
A. Loosely attached to the posterior abdominal wall
B. Almost entirely peritoneal
C. The widest part of the small intestine
D. The shortest segment of the small intestine
E. C-shaped Ref. 4 - pp. 1317-1318

655. THE SMALL INTESTINE:
A. Receives arterial blood from the inferior mesenteric artery
B. Is in part innervated by superior mesenteric plexus
C. Has a copious lymphatic plexus
D. All of the above
E. None of the above Ref. 4 - p. 1324

656. THE APPENDIX:
A. Does not have a true mesentery
B. Has a falciform fold of peritoneum
C. Has a variable position
D. Receives arterial blood from a branch of posterior cecal
E. None of the above Ref. 4 - pp. 1330-1332

657. THE FEMORAL RING IS BOUNDED:
A. Anteriorly by inguinal ligament
B. Medially by lacunar ligament
C. Laterally by femoral vein
D. Inferiorly by femoral artery
E. Posteriorly by fibers of Transversus abdominis
Ref. 4 - p. 547

658. INDIRECT INGUINAL HERNIAS:
A. Pass through deep inguinal ring
B. Most common hernia
C. Related to persistence of tunica vaginalis
D. More common in females
E. Accounts for 20 per cent of all hernias Ref. 4 - p. 547

659. THE LIVER IS:
A. The largest gland in the body
B. The most vascular gland in the body
C. Easily ruptured
D. Reddish-brown in living humans
E. Retroperitoneal Ref. 4 - pp. 1347-1348

660. THE LIVER IS ATTACHED BY FIVE PERITONEAL LIGAMENTS OR FOLDS:
A. Coronary
B. Falciform
C. Round
D. Triangular
E. All of the above Ref. 4 - pp. 1354-1355

SECTION V - THE ABDOMEN, PERINEUM AND PELVIS

661. THE INFERIOR POLE OF THE RIGHT KIDNEY LIES IN CONTACT WITH:
A. Duodenum
B. Jejunum
C. Inferior vena cava
D. Ovary
E. Transverse colon Ref. 4 - p. 1463

662. ANATOMICAL FEATURES OF THE KIDNEYS:
A. Immediately behind the peritoneum
B. Left lower than right
C. Attached primarily by peritoneum
D. Anterior surface of left is in direct contact with pancreas
E. Medial border of left closer to inferior vena cava
Ref. 4 - pp. 1460-1463

663. CONCERNING THE URETERS:
A. Average length 26 to 28 cm.
B. Crossed by spermatic or ovarian vessels
C. About 5 cm. apart when they reach bladder
D. Close to hypogastric artery at pelvic brim
E. None of the above Ref. 4 - pp. 1475-1476

664. THE AVERAGE URINARY BLADDER:
A. Can hold 1500 cc. of urine without overdistention
B. Is spherical when empty
C. Is pear-shaped in the child
D. Has a convex superior surface when distended
E. Is completely covered with peritoneum
Ref. 4 - pp. 1479-1480

665. THE BLOOD AND NERVE SUPPLY OF THE URINARY BLADDER:
A. Superior vesical artery
B. Branches of hypogastric artery
C. In female : branches of middle hemorrhoidal artery
D. Nerves from hypogastric sympathetic plexus
E. Parasympathetic fibers from vagus Ref. 4 - p. 1483

666. THE PROSTATE GLAND:
A. Consists only of glandular and connective tissues
B. Derives arterial supply from epigastric
C. Derives nerve supply from lumbar plexus
D. Has no lymphatics
E. None of the above Ref. 4 - pp. 1506-1507

667. CUTANEOUS NERVES OF THE SCROTUM:
A. Anterior scrotal branch of external pubogenital nerve
B. Superior scrotal branch of inferior epigastric nerve
C. Posterior scrotal branch of posterior femoral cutaneous nerve
D. Anterior scrotal branch of ilioinguinal nerve
E. Posterior scrotal branch of pudendal nerve
Ref. 5 - pp. 360-361

SECTION V - THE ABDOMEN, PERINEUM AND PELVIS

668. THE EXTERNAL ABDOMINAL OBLIQUE MUSCLE:
A. Is the most superficial of the anterolateral abdominal muscles
B. Arises from the lower borders of the lower eight ribs
C. Has a free posterior margin
D. Enters in the formation of the inguinal ligament
E. Is innervated by the inferior epigastric nerve
Ref. 5 - pp. 363-364

669. THE INTERNAL SPERMATIC FASCIA INVESTS THE:
A. Ductus deferens
B. Testicular vessels
C. Prostate gland
D. Testis
E. Seminal vesicle Ref. 5 - p. 370

670. A DIRECT INGUINAL HERNIA:
A. Is more common than an indirect hernia in males
B. Bulges and invariably descends into the scrotum
C. Is associated with the superior epigastric artery which lies lateral to it
D. Begins internally between the lateral margin of the falx inguinalis and the interfoveolar ligament
E. Is never covered with peritoneum Ref. 5 - p. 378

671. LAYERS COVERING A DIRECT INGUINAL HERNIA:
A. Transversalis fascia
B. Iliopectineal ligament and fascia
C. Cremaster muscle and fascial layer
D. Dartos fascia
E. Intercrural fibers of superficial inguinal ring
Ref. 5 - p. 378

672. DERIVATIVES OF THE DORSAL COMMON MESENTERY:
A. Transverse mesocolon
B. Falciform ligament
C. Sigmoid mesocolon
D. Greater omentum
E. Mesoappendix Ref. 5 - p. 380

673. THE SPLEEN:
A. Is yellow-green in color
B. Is related to the ninth, tenth and eleventh ribs
C. Is entirely extraperitoneal
D. Normally descends 5 cm below the costal margin
E. Receives nerves from the celiac plexus
Ref. 5 - pp. 400-401

674. THE PANCREAS:
A. Lies transversely across the abdominal wall from duodenum to spleen
B. Is behind the stomach
C. Has a main and an accessory duct
D. Develops in two parts from a dorsal and ventral primordium
E. Receives vessels from celiac trunk and superior mesenteric artery
Ref. 5 - pp. 405-406

SECTION V - THE ABDOMEN, PERINEUM AND PELVIS

675. THE GALL BLADDER:
A. Has a capacity of 250 cc
B. Is a diverticulum of the stomach
C. Is roughly pear-shaped
D. Lies free on the ventral hepatic surface
E. Does not come into contact with the peritoneum
Ref. 5 - p. 415

676. MECKEL'S DIVERTICULUM:
A. Occurs in about 25 per cent of the population
B. Is a finger-like pouch
C. Represents the remains of the urachus
D. Averages about 0. 5 cm in length
E. May be connected with the umbilical region
Ref. 5 - p. 421

677. THE SUPERIOR MESENTERIC ARTERY SUPPLIES THE:
A. Stomach
B. Proximal duodenum
C. Distal duodenum
D. Cecum
E. Ascending colon
Ref. 5 - p. 422

678. BRANCHES OF THE ILEOCOLIC ARTERY:
A. Posterior cecal
B. Appendicular
C. Ileal
D. Sigmoid
E. Rectal
Ref. 5 - p. 423

679. CHARACTERISTIC FEATURES OF THE VERMIFORM APPENDIX:
A. Average length about 8 cm
B. Located about 3 cm below the ileocecal valve
C. Most commonly tip lies retrocecally
D. No peritoneal investment
E. No lymphatic drainage
Ref. 5 - p. 428

680. BRANCHES OF THE INFERIOR MESENTERIC ARTERY:
A. Marginal
B. Left colic
C. Sigmoid
D. Superior rectal
E. Inferior pudendal
Ref. 5 - p. 430

681. ORGANS MAKING CONTACT WITH THE SURFACE OF THE LEFT KIDNEY:
A. Adrenal
B. Stomach
C. Pancreas
D. Cecum
E. Ovary
Ref. 5 - p. 434

SECTION V - THE ABDOMEN, PERINEUM AND PELVIS

682. THE LEFT SUPRARENAL GLAND IS IN RELATION WITH THE:
A. Cardia of the stomach
B. Spleen
C. Ascending colon
D. Liver
E. Pancreas Ref. 5 - p. 440

683. THE ANAL TRIANGLE OF THE PERINEUM IS BOUNDED BY THE:
A. Tip of the coccyx
B. Posterior border of the urogenital diaphragm
C. Sacrotuberous ligaments
D. Gluteus maximus muscle and piriformis muscle
E. External sphincter ani Ref. 5 - p. 471

684. LYMPHATICS OF THE URINARY BLADDER PASS TO THE:
A. Sacral nodes
B. Inguinal nodes
C. Internal iliac nodes
D. Popliteal nodes
E. External iliac nodes Ref. 5 - p. 482

685. ARTERIES OF THE PROSTATE GLAND ARE BRANCHES OF THE:
A. Posterior sacral
B. Middle rectal
C. Inferior vesical
D. Anterior ischial
E. Posterior perineal Ref. 5 - p. 488

ANSWER THE FOLLOWING QUESTIONS ACCORDING TO THE KEY BELOW:
A. If both statement and reason are true and related cause and effect
B. If both statement and reason are true but not related cause and effect
C. If the statement is true but the reason is false
D. If the statement is false but the reason is true
E. If both statement and reason are false

686. The ovarian artery reaches the ovary by way of the suspensory ligament
because
It is a branch of the abdominal aorta. Ref. 5 - p. 489

687. In the erect posture the uterus lies directly on the superior surface of the bladder
because
The uterus has no superior supports. Ref. 5 - p. 491

688. The medial umbilical ligament transmits urine from bladder to umbilicus
because
It is the obliterated umbilical artery. Ref. 5 - p. 499

689. The uterine artery is the homologue of the artery of the ductus deferens in the male
because
It is surrounded by endopelvic fascia. Ref. 5 - p. 500

SECTION V - THE ABDOMEN, PERINEUM AND PELVIS

690. The coccygeus muscle has no action
because
It has no osseous origin or insertion. Ref. 5 - p. 509

691. The recti muscles of the abdomen protect viscera from injury
because
When they contract they break the force of a blow.
Ref. 2 - p. 202

692. The pyramidalis muscle is commonly absent
because
It is inserted into the linea alba. Ref. 2 - p. 203

693. The lateral margin of the rectus abdominis muscle is practically a bloodless line
because
Very few branches of the epigastric artery cross it.
Ref. 2 - p. 204

694. In a malignant skin tumor arising below the level of the umbilicus it is important to first check the axillary lymph nodes
because
Superficial lymphatics of the entire anterior abdominal wall pass first to these nodes. Ref. 2 - p. 204

695. The internal oblique muscle has a less extensive aponeurosis than the external oblique muscle
because
It does not extend above the costal margin.
Ref. 2 - p. 206

696. The Dartos muscle of the scrotum is an involuntary muscle
because
It causes the skin of the scrotum to wrinkle when the outside temperature is low. Ref. 2 - p. 212

697. The ductus deferens conveys spermatozoa from the testis to the urethra
because
It is a continuation of the canal of the epididymis.
Ref. 2 - p. 213

698. Developmentally the falciform ligament is a mesentery which supports a blood vessel
because
During fetal life the umbilical artery is found along the free edge of this structure. Ref. 2 - pp. 227-228

699. The second part of the duodenum undergoes partial rotation to the right on its own long axis
because
Different parts of its wall grow unequally.
Ref. 2 - p. 262

SECTION V - THE ABDOMEN, PERINEUM AND PELVIS

700. The duodenum in man is not considered as part of the small intestine
because
It has lost its dorsal mesentery. Ref. 2 - p. 263

701. The easiest approach to the common bile duct is from the front
because
Only minor structures lie anterior to it.
Ref. 2 - p. 265

702. Ligation of the suprarenal artery from the aorta deprives the adrenal gland of its entire blood supply
because
Unlike most organs, the adrenal has no collateral arterial circulation.
Ref. 2 - p. 269

703. The lumbar vertebrae cannot be distinguished from thoracic vertebrae
because
Both have perforated transverse processes.
Ref. 2 - pp. 285-286

704. The linea semilunaris is a better site for surgical incisions than the linea alba
because
In the latter the nerves supplying the recti muscles may be cut.
Ref. 4 - p. 537

705. The inguinal canal is a true canal
because
In the female the ligament teres uteri passes through it.
Ref. 4 - p. 546

706. The diaphragm is more than a muscle of respiration
because
It aids in defecation and vomiting. Ref. 4 - p. 512

707. Direct inguinal hernias seldom become scrotal in position
because
They are most common in females. Ref. 4 - p. 547

708. The right lumbar arteries are longer than the left
because
The aorta lies a little to the left of the midline.
Ref. 4 - p. 733

709. The celiac trunk is the largest branch of the abdominal aorta
because
Its site of origin is close to the origin of the superior mesenteric artery.
Ref. 4 - p. 734

SECTION V - THE ABDOMEN, PERINEUM AND PELVIS

710. The collateral circulation, after ligation of the common iliac artery, is limited to the anastomosis of the external iliac and hypogastric communications

because

These are the two branches of the common iliac artery.

Ref. 4 - pp. 752-753

711. The lymph nodes of the hilus of the liver are of special clinical significance

because

When they are pathologically enlarged they compress the neck of the gallbladder. Ref. 4 - p. 888

712. In a patient with carcinoma of the pancreas, resection of the duodenal lymph nodes is not necessary

because

No anastomosis occurs between the lymphatics of these two organs.

Ref. 4 - p. 891

713. The sacral lymph nodes receive afferent vessels from the rectum and prostate gland

because

They are found in the perineum. Ref. 4 - p. 893

714. The epiploic foramen (of Winslow) is located caudal to the liver

because

It is the foramen into the cavity of the omentum.

Ref. 4 - p. 1298

715. The gastrocolic ligament is a portion of the lesser omentum

because

It extends from the greater curvature of the stomach to the transverse colon. Ref. 4 - p. 1297

SELECT THE ITEM FROM COLUMN I WHICH IS UNRELATED TO THE FOUR OTHER ITEMS IN THIS COLUMN, CIRCLE THIS LETTER. IDENTIFY THE ITEM IN COLUMN II TO WHICH THE FOUR ITEMS IN COLUMN I ARE RELATED, CIRCLE THIS NUMBER.

	COLUMN I	COLUMN II
716.	A. Gastrohepatic ligament	1. Portions of lesser omentum
	B. Gastrocolic ligament	2. Portions of greater omentum
	C. Gastrosplenic ligament	3. Ligaments supporting colon
	D. Gastrophrenic ligament	
	E. Lienorenal ligament	Ref. 4 - pp. 1297-1298
717.	A. Peritoneum	1. Coverings of femoral hernia
	B. Cremasteric fascia	2. Coverings of direct inguinal hernia
	C. Cribiform fascia	3. Coverings of umbilical hernia
	D. Tela subcutanea	
	E. Wall of femoral sheath	Ref. 4 - pp. 546-547

SECTION V - THE ABDOMEN, PERINEUM AND PELVIS

	COLUMN I	COLUMN II
718.	A. Double layer of peritoneum B. Ventral portion of primitive mesogastrium C. Cranial end continuous dorsally with coronary ligament D. Lower margin is free E. Represents obliterated fetal left umbilical vein	1. Falciform ligament 2. Round ligament of liver 3. Lesser omentum Ref. 4 - p. 1355
719.	A. Surrounded by pancreatic tissue B. About 75 cm. in length C. Spiral valve D. Accompanied by a few small lymph nodes E. Sphincter choledochus	1. Hepatic duct 2. Cystic duct 3. Common bile duct Ref. 4 - pp. 1309, 1373
720.	A. In contact with medial aspect of spleen B. Largely retroperitoneal C. Lies at level of 2nd and 3rd lumbar vertebrae D. Dorsal surface touches aorta E. Process uncinatus	1. Head of pancreas 2. Body of pancreas 3. Tail of pancreas Ref. 4 - pp. 1375-1377
721.	A. Superior mesenteric artery B. Ovarian artery C. Middle sacral artery D. Celiac artery E. Middle suprarenal artery	1. Visceral branches of abdominal aorta 2. Parietal branches of abdominal aorta 3. Paired branches of abdominal aorta Ref. 3 - p. 633
722.	A. Urethral artery B. Inferior rectal artery C. Deep artery of the penis D. Obturator artery E. Perineal artery	1. Arteries in Alcock's canal 2. Branches of internal pudendal artery 3. Branches of celiac trunk Ref. 3 - p. 651
723.	A. Inguinal ligament B. Pectineus covered by pectineal fascia C. Crescenteric base of the lacunar ligament D. Pubic tubercle E. Fibrous septum on medial side of femoral vein	1. Boundaries of the femoral triangle 2. Boundaries of the femoral sheath 3. Boundaries of the femoral ring Ref. 3 - p. 659

SECTION V - THE ABDOMEN, PERINEUM AND PELVIS

COLUMN I	COLUMN II

724.
A. Hepatic vein
B. Obturator vein
C. Middle rectal vein
D. Uterine vein
E. Vesical vein

1. Tributaries of internal iliac vein
2. Tributaries of inferior vena cava
3. Tributaries of portal vein

Ref. 3 - p. 717

725.
A. Splenic vein
B. Hepatic vein
C. Pyloric vein
D. Paraumbilical vein
E. Coronary vein

1. Tributaries of celiac vein
2. Tributaries of inferior vena cava
3. Tributaries of portal vein

Ref. 3 - p. 723

FOLLOWING THE CASE HISTORY ARE SEVERAL QUESTIONS BASED ON THE HISTORY. CHOOSE THE BEST ANSWER FOR EACH QUESTION.

A 52 year-old advertising executive, known to have had a bad duodenal ulcer for five years, is rushed to the hospital with massive vomiting of bright red blood. The patient had been "casually" following a regimen for peptic ulcer and had recently been under a "nervous strain." On admission to the hospital the patient reveals evidences of peritonitis, including absent bowel sounds and rigidity of the abdominal wall. Following appropriate supportive measures, including multiple blood transfusions; diagnostic studies are performed. It is ascertained that the patient most likely has a perforation of the first portion of the duodenum and surgery is performed.

726. AT SURGERY, ANTICIPATE FINDING THE FOLLOWING STRUCTURE(S) BEHIND THE FIRST PART OF THE DUODENUM:
A. Gastroduodenal artery
B. Common bile duct
C. Portal vein
D. All of the above
E. None of the above

Ref. 3 - p. 1229

727. WHICH OF THE FOLLOWING VESSELS ARE ASSOCIATED WITH THE BLOOD SUPPLY OF THE DUODENUM?:
A. Hepatic artery
B. Superior mesenteric artery
C. Superior mesenteric vein
D. Lienal vein
E. All of the above

Ref. 3 - p. 1230

728. IN FREEING THE DUODENUM, THE DUODENO-JEJUNAL FLEXURE IS HELD IN PLACE BY THE:
A. Falciform ligament
B. Ligament of Treitz
C. Greater omentum
D. Lesser omentum
E. Hepatoduodenal ligament

Ref. 3 - p. 1230

SECTION V - THE ABDOMEN, PERINEUM AND PELVIS

729. WHICH PART OF THE DUODENUM IS FOUND ON THE LEFT SIDE OF THE AORTA, IN FRONT OF THE LEFT RENAL VESSELS AND COVERED IN FRONT BY PERITONEUM CONTINUOUS WITH LEFT PORTION OF MESENTERY?:
A. Superior (1st portion)
B. Descending (2nd portion)
C. Horizontal (3rd portion)
D. Ascending (4th portion)
E. None of the above Ref. 3 - p. 1230

730. IF BIOPSY REVEALS MALIGNANT CELLS IN THE REGION OF THE PERFORATION, LOOK FOR LYMPH NODE INVOLVEMENT IN THE AREA OF THE:
A. Superior mesenteric artery
B. Spleen
C. Inguinal lymph nodes
D. Axillary lymph nodes
E. Inferior mesenteric artery Ref. 3 - p. 763

SECTION V – THE ABDOMEN, PERINEUM AND PELVIS

IDENTIFY THE COVERINGS AND STRUCTURES ASSOCIATED WITH THE INGUINAL CANAL AND SCROTUM.

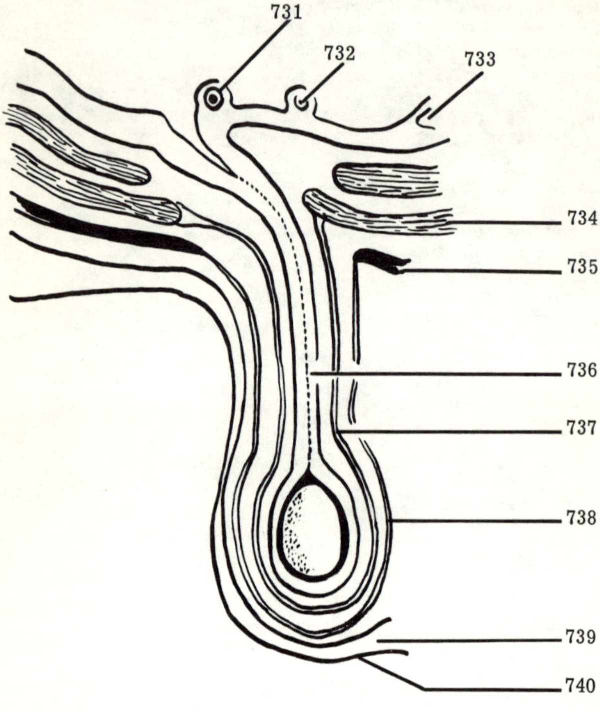

A. Bladder
B. Colles' fascia
C. External oblique aponeurosis
D. Internal oblique
E. Vas deferens
F. Inferior epigastric artery
G. Rectus sheath
H. Transversus
I. Inguinal canal
J. Peritoneum
K. Cavity of tunica vaginalis
L. Obliterated umbilical artery
M. Inguinal ligament
N. Prostate gland
O. Skin

Ref. 1 - p. 113

SECTION V - THE ABDOMEN, PERINEUM AND PELVIS

IDENTIFY THE BRANCHES OF DISTRIBUTION OF THE CELIAC ARTERY.

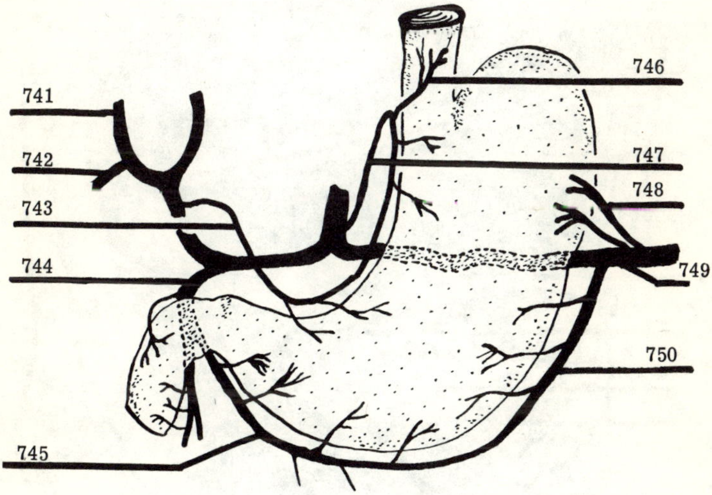

A. Right hepatic
B. Cystic
C. Right gastric
D. Gastro-duodenal
E. Right gastro-epiploic
F. Esophageal
G. Left gastric
H. Short gastric
I. Splenic
J. Left gastro-epiploic
K. Left hepatic
L. Supra-duodenal
M. Superior pancreatico-duodenal
N. Phrenic
O. Inferior pancreatic

Ref. 1 - p. 132.1

SECTION V - THE ABDOMEN, PERINEUM AND PELVIS

IDENTIFY THE STRUCTURES IN THE MEDIAN SECTION THROUGH THE FEMALE PELVIS.

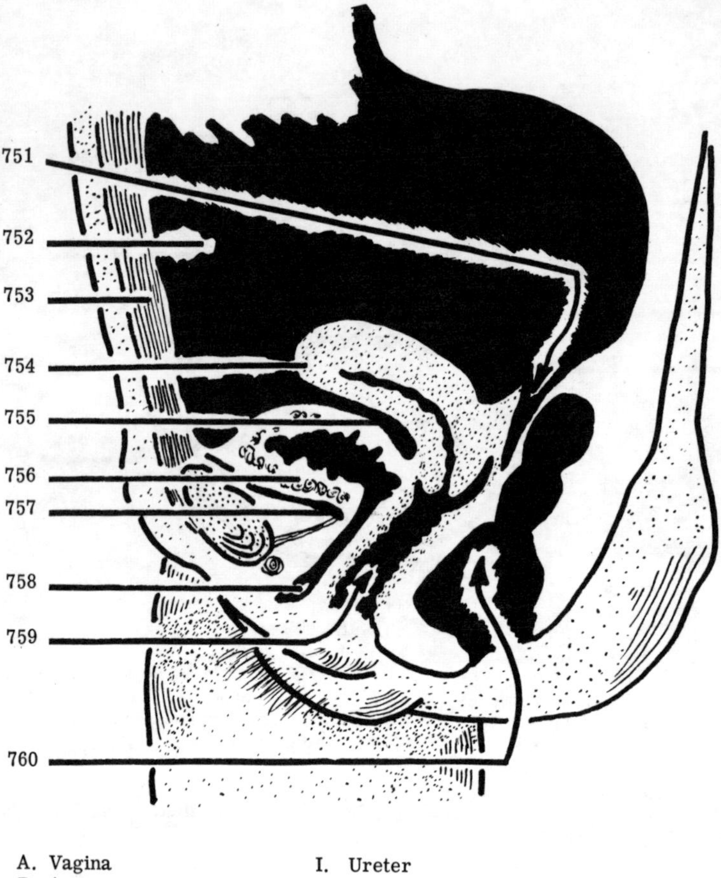

A. Vagina
B. Anus
C. Recto-uterine pouch
D. Rectum
E. Rectus abdominis muscle
F. Urethra
G. Uterus
H. Retro-pubic space
I. Ureter
J. Bladder
K. Levator ani muscle
L. Pleural cavity
M. Utero-vesical pouch
N. Cervix
O. Peritoneal cavity

Ref. 1 - p. 236

FOR EACH OF THE FOLLOWING MULTIPLE CHOICE QUESTIONS SELECT THE ONE MOST APPROPRIATE ANSWER:

761. THE CELL THEORY WAS FORMULATED BY:
A. Leeuwenhoek
B. Hooke
C. Virchow
D. Schleiden and Schwann
E. Darwin and Amici — Ref. 6 - p. 1

762. THE CONCEPT THAT HUMAN DISEASE CAN BE TRACED TO ALTERATIONS IN THE CELLS IS ATTRIBUTED TO:
A. Cajal
B. Virchow
C. Flemming
D. Darwin
E. Hooke — Ref. 6 - p. 1

763. THE TECHNIQUE OF TISSUE CULTURE WAS FIRST UTILIZED BY:
A. Harrison
B. Flemming
C. Sabin
D. Enders
E. Golgi — Ref. 6 - p. 3

764. VITAL STAINING WITH JANUS GREEN IS USEFUL IN STUDYING:
A. Lysosomes
B. Golgi apparatus
C. Spindle fibers
D. Mitochondria
E. Nucleoli — Ref. 6 - p. 7

765. GLUCOSE-6-PHOSPHATE IS OFTEN USED AS AN IDENTIFYING MARKER IN CENTRIFUGAL FRACTIONS FOR:
A. Microsomes
B. Centrosomes
C. Nucleoli
D. Mitochondria
E. Lysosomes — Ref. 6 - p. 8

766. THE STAIN OF CHOICE TO DEMONSTRATE FATTY COMPONENTS IN A CELL IS:
A. Eosin
B. PAS
C. Sudan black
D. Toluidine blue
E. Hematoxylin — Ref. 6 - p. 11

767. THE FEULGEN REACTION IS A SPECIFIC HISTOCHEMICAL METHOD FOR DEMONSTRATING:
A. RNA
B. DNA
C. Glycogen
D. Lipids
E. Glucose-6-phosphate — Ref. 6 - p. 17

768. THE FLUORESCENT ANTIBODY TECHNIQUE HAS BEEN USED FOR LOCALIZATION OF SPECIFIC __________ IN TISSUES:
A. Proteins
B. Carbohydrates
C. Lipids
D. Parasites
E. Bacteria Ref. 6 - p. 19

769. THE OVERALL RESOLVING POWER OF A GOOD LIGHT MICROSCOPE, OF HIGH NUMERICAL APERTURE, IS APPROXIMATELY:
A. 0.005 u
B. 0.25 u
C. 0.50 u
D. 0.85 u
E. 1.0 u Ref. 6 - p. 21

770. LIVING CELLS ARE BEST STUDIED WITH THE:
A. Ultracentrifuge
B. Electron miscroscope
C. Light microscope
D. Phase microscope
E. Flame spectrophotometer Ref. 6 - p. 23

771. THE OVERALL LIMIT OF RESOLUTION OF THE ELECTRON MICROSCOPE, FOR BIOLOGICAL PREPARATIONS, IS ABOUT:
A. 1 Å
B. 5 Å
C. 15 Å
D. 50 Å
E. 100 Å Ref. 6 - p. 25

772. THE MOST WIDELY USED FIXATIVE FOR ELECTRON MISCROCOPE PREPARATIONS UTILIZES:
A. Lead
B. Zinc
C. Osmium
D. Platinum
E. Mercury Ref. 6 - p. 26

773. OUTSIDE OF THE CELL, ONE OF THE MOST COMMON INORGANIC CONSTITUENTS IS:
A. Sodium
B. Potassium
C. Calcium
D. Magnesium
E. Iron Ref. 6 - p. 30

774. HYALURONIC ACID IS USUALLY NOT ASSOCIATED WITH:
A. Cartilage
B. Aqueous humor of the eye
C. Fluids of the chest
D. Joint cavities
E. Duodenal secretions Ref. 6 - p. 31

775. DNA IS FOUND ONLY IN THE:
A. Nucleus
B. Mitochondria
C. Golgi apparatus
D. Lysosomes
E. Ribosomes Ref. 6 - p. 33

776. ADHERING TO THE OUTER SURFACE OF THE LIMITING MEMBRANE OF THE ENDOPLASMIC RETICULUM ARE:
A. Mitochondria
B. Lysosomes
C. Ribosomes
D. Golgi particles
E. Vacuoles Ref. 6 - p. 41

777. THE ORGANELLES ACTING AS POWER SOURCES, SUPPLYING ENERGY FOR NUMEROUS CHEMICAL REACTIONS IN THE CELL ARE:
A. Mitochondria
B. Lysosomes
C. Ribosomes
D. Golgi particles
E. Vacuoles Ref. 6 - p. 44

778. THE ORGANELLES PLAYING A PART IN THE INTRACELLULAR DIGESTION OF PHAGOCYTOSED PARTICLES ARE CALLED:
A. Endocytes
B. Lysosomes
C. Vacuolocytes
D. Mitochondria
E. Centrosomes Ref. 6 - p. 47

779. WHICH OF THE FOLLOWING IS USED AS A DIFFERENTIAL STAIN FOR DISTINGUISHING RNA FROM DNA?:
A. Hematoxylin
B. Feulgen stain
C. 3 per cent orcein
D. Janus green B
E. Eosin Ref. 6 - p. 55

780. BEFORE MITOSIS CAN BE OPTICALLY OBSERVED, THE PROCESS OF ____________ HAS OCCURRED:
A. Chromosomal replication
B. Splitting of the centrosome
C. Disolution of ribosomes
D. Mitochondrial degradation
E. Vacuolization Ref. 6 - p. 61

781. CHROMATIDS ARE FORMED DURING:
A. Interphase
B. Prophase
C. Metaphase
D. Telophase
E. Anaphase Ref. 6 - p. 62

782. THE HUMAN MALE HAS A KARYOTYPE OF:
A. 20 pairs of autosomes and an X and a Y
B. 22 pairs of autosomes and two X's
C. 22 pairs of autosomes and an X and a Y
D. 23 pairs of autosomes and two Y's
E. 24 pairs of autosomes and an X and a Y
Ref. 6 - p. 71

783. ONE TYPE OF EPITHELIUM MAY CHANGE INTO ANOTHER BY A PROCESS CALLED:
A. Regeneration
B. Amitosis
C. Meiosis
D. Pinocytosis
E. Metaplasia
Ref. 6 - p. 84

784. THE STRIATED OR BRUSH BORDER IS COMPOSED OF:
A. Tightly packed cilia
B. Microvilli
C. Tightly packed filaments
D. Dendrites
E. Tonofibrils
Ref. 6 - p. 89

785. AT THE BASE OF EACH CILIUM IS A DENSE GRANULE CALLED THE:
A. Centrosome
B. Nissl body
C. Basal body
D. Goblet cell
E. Macrosome
Ref. 6 - p. 91

FOR EACH OF THE FOLLOWING MULTIPLE CHOICE QUESTIONS SELECT THE APPROPRIATE ANSWER OR ANSWERS. ONE OR MORE, OR ALL, OF THE CHOICES MAY BE USED IN CORRECTLY ANSWERING THE QUESTION:

786. EPITHELIUM IS CHARACTERIZED BY:
A. Small amounts of intercellular substance
B. Large cells with small nucleoli
C. Cells devoid of mitochondria
D. Lack of a basement membrane
E. Location : lining walls of internal cavities
Ref. 6 - pp. 77-93

787. SIMPLE CUBOIDAL EPITHELIUM IS USUALLY FOUND IN:
A. The inner surface of the tympanic membrane of the ear
B. The parietal layer of Bowman's capsule
C. The descending limb of the loop of Henle
D. The rete testis
E. A single layer on the free surface of the ovary
Ref. 6 - p. 80

788. THE CENTRAL CANAL OF THE SPINAL CORD IS LINED BY:
A. Simple squamous epithelium
B. Simple cuboidal epithelium
C. Simple columnar epithelium
D. Simple columnar ciliated epithelium
E. None of the above Ref. 6 - p. 82

789. THE EUSTACHIAN TUBE IS LINED BY:
A. Simple ciliated epithelium
B. Simple columnar epithelium
C. Simple cuboidal epithelium
D. Ciliated simple columnar epithelium
E. Ciliated pseudostratified columnar epithelium
Ref. 6 - p. 83

790. TRANSITIONAL EPITHELIUM IS FOUND IN THE:
A. Ureter
B. Urethra
C. Urinary bladder
D. Liver
E. Duodenum Ref. 6 - p. 84

791. IN MAMMALS, ERYTHROCYTES ARE:
A. Nonmotile
B. Without a nucleus
C. Devoid of mitochondria
D. Pale, greenish yellow
E. Carriers of oxygen Ref. 6 - pp. 116-117

792. IN MAN THE LYMPHOCYTES:
A. Form 40 to 45 per cent of the total number of colorless corpuscles
B. Are spherical and slightly larger than erythrocytes
C. Are nongranular leukocytes
D. Produce fibrinogen
E. Require vital stains to demonstrate azurophilic granules
Ref. 6 - p. 119

793. ACIDOPHILIC OR EOSINOPHILIC LEUKOCYTES:
A. Are about 9 u in diameter, in the fresh condition
B. Vary from 2 to 5 per cent of total leukocyte count
C. Usually have a small pyknotic nucleus
D. Generally have a prominent nucleolus
E. Have nuclei that stain red Ref. 6 - p. 122

794. MICROSCOPIC SECTIONS OF TENDONS REVEAL:
A. Parallel bundles of elastic fibers
B. Chief cellular elements are fibroblasts
C. Numerous histiocytes and mast cells
D. No differences from sections of ligaments
E. All of the above Ref. 6 - p. 156

795. THE MATRIX OF HYALINE CARTILAGE IS:
A. Markedly acidophilic
B. Composed of chondromucoprotein
C. Devoid of blood vessels
D. Penetrated by thin collagenous fibrils
E. Able to regenerate independently Ref. 6 - p. 214

796. VERTEBRATE CARDIAC MUSCLE IS CHARACTERIZED BY:
A. A syncytium
B. Simple cylindrical units
C. Nuclei are situated at periphery of muscle cell
D. No sarcolemma
E. Rhythmical contraction Ref. 6 - p. 285

797. GOLGI TYPE II NEURONS ARE ABUNDANT IN THE:
A. Cerebral cortex
B. Pons
C. Cerebellar cortex
D. Spinal cord
E. Retina Ref. 6 - p. 317

798. TYPES OF NEUROGLIA:
A. Astrocytes
B. Microglia
C. Plasmocytes
D. Fibrocytes
E. Oligodendrocytes Ref. 6 - p. 335

799. IN LARGE ELASTIC ARTERIES THE:
A. Tunica intima is about 25 u thick
B. Endothelium is composed of transitional epithelium
C. Tunica media consists largely of elastic tissue
D. Tunica adventitia is very thick
E. Walls may contain smooth muscle fibers
Ref. 6 - pp. 368-369

800. SINUSOIDS DIFFER FROM CAPILLARIES IN THAT THE FORMER HAVE:
A. Irregular tortuous walls
B. A continuous layer of endothelial cells
C. A dense connective tissue wall
D. A wide distribution in all organs
E. More phagocytic elements Ref. 6 - pp. 362-363

801. MUSCULAR ARTERIES (MEDIUM CALIBER) EXHIBIT:
A. A well-developed internal elastic membrane
B. A tunica media composed of elastic fibers
C. A tunica adventitia of loose connective tissue
D. A tunica intima lined by mesothelium
E. All of the above Ref. 6 - p. 368

802. CONCERNING THE HISTOLOGICAL STRUCTURE OF THE THYMUS:
A. Medulla contains reticular cells
B. Medulla contains Hassal bodies
C. Medulla more vascular than cortex
D. Cells identical to lymphocytes in cortex
E. All of the above Ref. 6 - pp. 417-421

803. THE EPIDERMAL STRUCTURE OF THE PALMS AND SOLES REVEAL:
A. Thicker Malpighian layer between dermal papillae
B. Keratohyalin in Malpighian layer
C. Eleidin is stratum lucidum
D. Thick stratum corneum
E. Keratin in stratum corneum Ref. 6 - pp. 481-483

804. MICROSCOPIC SECTIONS OF HAIR SHOW:
A. A special membrane : the cuticle
B. One or more sebaceous glands associated with the hair
C. A main mass of dense, horny substance
D. None of the above
E. All of the above Ref. 6 - pp. 491-494

805. THE PAROTID GLAND IS CHARACTERIZED BY:
A. Albuminous (serous) secretions
B. Numerous striated tubules
C. Secretory portions have form of "grapes"
D. In newborn : a positive staining reaction for mucous
E. All of the above Ref. 5 - pp. 402-404

806. THE PERIODONTAL MEMBRANE FULFILLS THE FOLLOWING CRITERIA:
A. Serves as a periosteum to alveolar bone
B. Connects the root and the crown of the tooth
C. Is rich in elastic fibers
D. All of the above
E. None of the above Ref. 6 - p. 537

807. IDENTIFYING FEATURES OF THE ESOPHAGUS INCLUDE:
A. A connective tissue tube
B. Epithelium of simple columnar type
C. Lamina propria of dense connective tissue
D. No tunica adventitia
E. Presence of cardiac glands Ref. 6 - pp. 544-546

808. CELLS CHARACTERISTICALLY FOUND IN THE GASTRIC GLANDS:
A. Zymogenic cells
B. Pseudostratified columnar cells
C. Argentaffine cells
D. Mucous neck cells
E. Parietal cells Ref. 6 - p. 550

809. CONCERNING THE LAMINA PROPRIA OF THE LARGE INTESTINE:
A. Structure same as in the small intestine
B. Abundance of eosinophiles
C. Scattered solitary follicles
D. None of the above
E. All of the above Ref. 6 - pp. 569, 571

810. RELATIVE TO THE STRUCTURE OF THE LIVER LOBULE:
A. Arranged as a polygonal prism
B. Central artery runs through center of lobule
C. At periphery of lobule are branches of portal vein
D. In man there are well-developed connective tissue partitions
E. Height of lobule greater than cross section
Ref. 6 - pp. 583-584

ANSWER THE FOLLOWING QUESTIONS BY USING THE KEY AS OUTLINED BELOW:

A. If 1, 2 and 3 are correct
B. If 1 and 3 are correct
C. If 2 and 4 are correct
D. If all are correct
E. If some combination other than the above are correct
F. If all are incorrect

811. THE SPLEEN:
1. Has a dense connective tissue capsule containing a few smooth muscle cells
2. Contains Billroth cords
3. Cannot undergo myeloid metaplasia
4. Contains muscular arteries of medium caliber

Ref. 6 - pp. 403-407

812. CELLS ASSOCIATED WITH THE THYMUS:
1. Lymphocytes
2. Epithelial reticular cells
3. Macrophages
4. Paneth cells

Ref. 6 - pp. 417-418

813. ACIDOPHILS (ALPHA CELLS) OF THE HYPOPHYSIS ARE BELIEVED TO SECRETE:
1. Thyrotropic hormone
2. Growth hormone
3. Follicle stimulating hormone
4. Prolactin (LTH)

Ref. 6 - p. 431

814. TYPES OF EPITHELIAL CELLS ASSOCIATED WITH THE PARATHYROID GLANDS:
1. Beta cells
2. Chief cells
3. Hassal bodies
4. Oxyphilic cells

Ref. 6 - p. 454

815. THE CHROMAFFIN CELLS OF THE ADRENAL GLAND:
1. Are derived from neuroectoderm
2. Secrete catecholamines
3. Are innervated by preganglionic sympathetic fibers
4. Are localized to the cortex

Ref. 6 - p. 460

816. FOLLOWING ABLATION OF THE ADRENAL CORTEX THERE IS:
1. Excessive excretion of sodium in the urine
2. Plasma potassium is lowered
3. Hypotension
4. Tissue dehydration

Ref. 6 - p. 463

817. THE PINEAL BODY:
1. Is invested by pia mater
2. Contains chief cells
3. Contains interstitial cells
4. Contains corpora arenacea

Ref. 6 - pp. 472,474

818. THE STRATUM MALPIGHII OF THE EPIDERMIS IS SUBDIVIDED INTO THE:
1. Stratum germinativum
2. Stratum granulosum
3. Stratum spinosum
4. Stratum corneum Ref. 6 - p. 481

819. THE COLOR OF THE SKIN IS THE RESULT OF:
1. Carotene
2. Oxyhemoglobin
3. Melanin
4. Biliverdin Ref. 6 - p. 486

820. THE SEBACEOUS GLANDS:
1. Lie in the dermis
2. Are not found in the palms and soles
3. Are of the holocrine type
4. Have excretory ducts that open into necks of hair follicles
Ref. 6 - p. 497

821. SWEAT GLANDS ARE:
1. Simple, coiled, tubular glands
2. Apocrine type in axilla
3. Not found along lip margins
4. Noted to have twisted, spiral excretory ducts in the epidermis
Ref. 6 - pp. 497,499

822. BASAL (BASKET) CELLS ARE ASSOCIATED WITH THE:
1. Parotid gland
2. Gastric glands
3. Thyroid gland
4. Medium size muscular arteries Ref. 6 - p. 519

823. CELLS FOUND IN THE PALATINE TONSIL:
1. Plasma cell
2. Lymphocyte
3. Epithelial cell
4. Neutrophilic granular leukocyte Ref. 6 - p. 525

824. CELLS ASSOCIATED WITH THE CEMENTUM OF THE TOOTH:
1. Odontoblasts
2. Ameloblasts
3. Osteocytes
4. Tomocytes Ref. 6 - p. 535

825. THE ESOPHAGEAL GLANDS PROPER:
1. Are small, compound glands
2. Contain only mucous cells
3. Are found in the submucous layer
4. Have ducts lined by simple ciliated cuboidal epithelium
Ref. 6 - p. 545

826. GASTRIC GLANDS OF THE STOMACH CONTAIN:
1. Zymogenic cells
2. Parietal cells
3. Alpha cells
4. Argentaffin cells Ref. 6 - p. 550

827. ARGENTAFFIN CELLS OF THE STOMACH:
1. Are found lining the surface of the epithelium
2. Contain no nucleus
3. Stain specifically with Feulgen stain
4. Store hydrochloric acid Ref. 6 - p. 553

828. PANETH CELLS:
1. Are found in the glands of Lieberkühn
2. Produce pepsinogen
3. Have intensely basophilic basal cytoplasm
4. Are multinucleated Ref. 6 - pp. 564-565

829. PEYER'S PATCHES:
1. As a rule, occur only in the jejunum
2. Always occur on the intestinal wall opposite the mesentery attachment
3. Hypertrophy in old age
4. Consist of dense lymphatic tissue Ref. 6 - p. 567

830. CANALS OF HERING ARE FOUND IN THE:
1. Spleen
2. Liver
3. Pancreas
4. Appendix Ref. 6 - p. 585

831. HEPATIC SINUSOIDS:
1. Are larger than capillaries
2. Are irregular in shape
3. Are not surrounded by connective tissue
4. Have phagocytic cells in their walls Ref. 6 - p. 588

832. THE WALL OF THE GALLBLADDER CONSISTS OF THE FOLLOWING LAYERS:
1. Mucous layer
2. Perimuscular connective tissue layer
3. Serous layer
4. Adventitial layer Ref. 6 - p. 605

833. THE ISLETS OF LANGERHANS:
1. Are richly vascularized
2. Range from 200,000 to 1,800,000 in the adult human
3. Surrounded by a thin layer of reticular fibers
4. Contain alpha, beta and delta cells Ref. 6 - pp. 616-617

834. THE PROXIMAL TUBULE OF THE KIDNEY:
1. Is extremely tortuous
2. Makes up the bulk of the kidney medulla
3. Is highly branched
4. Continues into the loop of Henle Ref. 6 - p. 657

835. JUXTAGLOMERULAR CELLS:
1. Probably produce renin
2. Are in contact with the intima of the afferent arteriole
3. Have a basophilic cytoplasm
4. Have specific granules clearly demonstrated by the Bowie stain
Ref. 6 - p. 665

FOR THE FOLLOWING GROUP OF QUESTIONS ANSWER ACCORDING TO THE KEY OUTLINED BELOW:
A. If only A is correct
B. If only B is correct
C. If A and B are correct
D. If A and B are incorrect

836. HISTOPHYSIOLOGY OF THE KIDNEYS REVEALS THAT THE:
A. Hydrostatic pressure in glomerular capillaries is about 70 mm./Hg.
B. Glomerulus has a filtration pressure (net) of about 18 mm./Hg.
C. Both
D. Neither
Ref. 6 - p. 669

837. THE COLLICULUS SEMINALIS:
A. Is an elevation on the posterior wall of the male urethra
B. Has an opening in its midline for epididymis
C. Both
D. Neither
Ref. 6 - p. 680

838. THE SURFACE OF THE MUCOUS MEMBRANE OF THE URETHRA SHOWS MANY RECESSES DUE TO:
A. Lacunae of Morgagni
B. Glands of Bartholin
C. Both
D. Neither
Ref. 6 - p. 680

839. ENDOCRINE TISSUE OF THE TESTIS:
A. Sertoli cells
B. Leydig cels
C. Both
D. Neither
Ref. 6 - p. 685

840. THE DUCTUS EPIDIDYMIS:
A. Is highly tortuous
B. Is lined by stereocilia
C. Both
D. Neither
Ref. 6 - p. 710

841. THE EJACULATORY DUCTS:
A. Are lined with ciliated simple columnar epithelium
B. Are branched into glands of Littre
C. Both
D. Neither
Ref. 6 - pp. 713-714

842. THE BULBOURETHRAL GLANDS ARE:
A. Also called glands of Bartholin
B. In some respects similar to mucous glands
C. Both
D. Neither
Ref. 6 - p. 719

843. THE OVARIAN FOLLICLES:
A. Are found primarily in the medulla of the ovary
B. Average about 50,000 in a normal young adult
C. Both
D. Neither
Ref. 6 - p. 728

844. THE CORPUS LUTEUM SECRETES:
A. Progesterone
B. Prolactin
C. Both
D. Neither
Ref. 6 - p. 737

845. WALLS OF THE OVIDUCT:
A. Submucous layer
B. Mucous membrane
C. Both
D. Neither
Ref. 6 - pp. 743-744

846. THE MUCOUS MEMBRANE OF THE VAGINA CONSISTS OF:
A. Stratified squamous epithelium
B. Lamina propria
C. Both
D. Neither
Ref. 6 - p. 763

847. THE SKIN OF THE NIPPLE AND AREOLA IS RICH IN:
A. Free nerve endings
B. Meissner's corpuscles
C. Both
D. Neither
Ref. 6 - p. 769

848. THE SCLERA OF THE EYE IS:
A. Rich in reticular fibers
B. Associated with flat, elongated fibroblasts
C. Both
D. Neither
Ref. 6 - p. 780

849. BOWMAN'S MEMBRANE IS:
A. Associated with the cornea
B. Composed of elastic fibers
C. Both
D. Neither
Ref. 6 - p. 781

850. THE ROD CELLS OF THE RETINA ARE:
A. Associated with visual purple
B. Particularly concentrated in the fovea
C. Both
D. Neither
Ref. 6 - p. 793

851. THE GLANDS OF MOLL ARE:
A. Peculiar sweat glands
B. Found in the eyelids
C. Both
D. Neither
Ref. 6 - p. 807

852. THE LACRIMAL GLAND:
A. Is of the tubuloalveolar type
B. Secretes mucous
C. Both
D. Neither Ref. 6 - p. 809

853. PSEUDOSTRATIFIED COLUMNAR EPITHELIUM IS FOUND IN THE:
A. Male urethra
B. Vagina
C. Both
D. Neither Ref. 6 - p. 83

854. STEREOCILIA IS FOUND IN THE EPITHELIUM LINING THE:
A. Epididymis
B. Trachea
C. Both
D. Neither Ref. 6 - p. 90

855. ANISOCYTOSIS IMPLIES THAT ERYTHROCYTES ARE:
A. Unusually large
B. Unusually small
C. Both
D. Neither Ref. 6 - p. 116

856. CHARACTERISTIC OF MONOCYTES:
A. Large cells, 9 to 12 microns
B. Oval or kidney-shaped nucleus
C. Both
D. Neither Ref. 6 - p. 121

857. COLLAGENOUS FIBERS ARE DIGESTED BY:
A. Alkaline trypsin
B. Pepsin in acid solutions
C. Both
D. Neither Ref. 6 - p. 133

858. IDENTIFYING FEATURES OF MAST CELLS:
A. Often in groups about blood vessels
B. Granules have purple color with methylene blue stain
C. Both
D. Neither Ref. 6 - p. 152

859. ELASTIC CARTILAGE IS FOUND IN THE:
A. Ventral ends of ribs
B. Respiratory passages
C. Both
D. Neither Ref. 6 - p. 216

860. CALCIFICATION AND ASBESTOS FORMATION MAY OCCUR IN:
A. Hyaline cartilage
B. Elastic cartilage
C. Both
D. Neither Ref. 6 - p. 219

ANSWER THE FOLLOWING QUESTIONS ACCORDING TO THE KEY OUTLINED BELOW:

A. If both statement and reason are true and related cause and effect
B. If both statement and reason are true but not related cause and effect
C. If the statement is true but the reason is false
D. If statement is false and reason is true
E. If both statement and reason are false

861. Neutral red and Janus green have been widely used in the study of living blood leukocytes

because

They are of relatively low toxicity and can be used to study living cells.
Ref. 6 - pp. 6-7

862. Hematoxylin and eosin is the most commonly used staining method

because

It does not require fixation of tissues. Ref. 6 - p. 10

863. Except for enzymes, chemical methods are available for study cells

because

The enzymes digest the fixatives and stains.
Ref. 6 - pp. 16-17

864. The cytoplasm of nearly all cells is homogeneous except for a few embedded organelles

because

The endoplasmic reticulum was discovered by using the electron microscope.
Ref. 6 - p. 41

865. The Golgi apparatus is readily studied with vital stains

because

It has the same morphology when viewed either by light or electron microscopes. Ref. 6 - p. 48

866. Pseudostratified columnar epithelium is really a modification of simple columnar epithelium

because

Although the nuclei appear to be at different levels, the basal portions of the cell touch the basement membrane.
Ref. 6 - p. 79

867. Stratified squamous epithelium is found lining the cornea

because

It is a type of epithelium that can become keratinized.
Ref. 6 - p. 82

868. Tonofibrils are common in epithelial cells

because

They pass from one cell body to another through intercellular bridges.
Ref. 6 - pp. 84-86

869. The basal lamina is a condensation of connective tissue ground substance

because

It seems to be composed of a collagen. Ref. 6 - p. 89

870. The highest structural differentiation of the free surface of the epithelium is reached in ciliated cells
because
Tonofibrils connect at the basal granule and integrate the wave of motion of the cilia. Ref. 6 - pp. 91-93

871. As a rule epithelial tissue is devoid of blood vessels
because
Epithelial cells do not require oxygen. Ref. 6 - p. 93

872. The goblet cell is a descriptive form of the cell
because
It is usually found scattered among columnar cells.
Ref. 6 - p. 101

873. In simple tubular glands the excretory duct opens as a pore
because
These glands only secrete mucous.
Ref. 6 - p. 104

874. Blood and lymph are not classified as connective tissue
because
The intercellular substance is liquid. Ref. 6 - p. 111

875. In mammals, erythrocytes are non-living cells
because
In all other vertebrates they are found to have a nucleus
Ref. 6 - p. 112

876. The correct term for lymphocytes is "azurophils"
because
The azurophilic granules are a constant feature of this cell.
Ref. 6 - p. 119

877 Elastic fibers are abundant in loose connective tissue
because
These fibers are not fibrillar but are usually homogeneous.
Ref. 6 - pp. 133-135

878. Sinusoids of bone marrow have a distinct epithelial cell wall
because
Sinusoids are found only in the bone marrow.
Ref. 6 - p. 184

879. Longitudinal striations are very pronounced in cardiac muscle
because
The sarcoplasm is usually abundant in cardiac muscle.
Ref. 6 - pp. 286-287

880. Schwann's cells are necessary for the life and function of the axon of the peripheral nerves
because
They are the chief factor in the normal metabolism of the axon.
Ref. 6 - pp. 321-322

881. The only component of the wall of a capillary is the endothelium
because
Endothelial cells are structurally very similar to fibroblasts.
Ref. 6 - pp. 358-359

882. The follicles of the normal human thyroid are lined by a low cuboidal epithelium
because
There is no distinct basement membrane.
Ref. 6 - pp. 445-446

883. When the tongue is anesthetized all taste is lost
because
Taste buds are found only on the tongue. Ref. 6 - p. 514

884. Basal (basket) cells are presumed to be contractile
because
They resemble myoepithelial cells of the sweat glands.
Ref. 6 - p. 519

885. Dentin of the tooth is sensitive to touch and to cold
because
It is rich and complexly penetrated with nerve fibers.
Ref. 6 - p. 531

ANSWER THE FOLLOWING QUESTIONS BY USING THE KEY OUTLINED BELOW:
A. If A is greater in severity, frequency or size than B
B. If B is greater in severity, frequency or size than A
C. If A and B are equal in severity, frequency or size

886. TENDENCY OF CELLS TO ADHERE TO ONE ANOTHER:
A. Erythrocytes
B. Leukocytes Ref. 6 - p. 112

887. RELATIVE AMOUNTS OF CYTOPLASM:
A. Lymphocytes
B. Monocytes Ref. 6 - p. 121

888. CHARACTERISTIC CONSTITUENT OF ALL TYPES OF CONNECTIVE TISSUE:
A. Collagenous fibers
B. Elastic fibers Ref. 6 - p. 132

889. CELLS IN LOOSE CONNECTIVE TISSUE:
A. Macrophages (histiocytes)
B. Lymphocytes Ref. 6 - p. 146

890. LYMPHOCYTES IN LYMPHATIC TISSUE:
A. Small
B. Large Ref. 6 - p. 145

891. TYPES OF NEURONS IN THE NERVOUS SYSTEM:
A. Bipolar
B. Multipolar Ref. 6 - p. 318

892. RATE OF CONDUCTION OF IMPULSES:
A. Large axons
B. Small axons Ref. 6 - p. 325

893. DISTANCE FROM THE SKIN SURFACE:
A. Lymphatic networks
B. Blood capillary networks Ref. 6 - p. 390

894. SIZE OF THE RED PULP OF THE SPLEEN IN:
A. Lymphatic leukemia
B. Myeloid leukemia Ref. 6 - p. 405

895. LYMPHATICS OF THE ADRENAL GLAND:
A. Zona fasciculata
B. Zona reticularis Ref. 6 - p. 463

896. EFFECT OF NOREPINEPHRINE ON:
A. Blood pressure
B. Cardiac output Ref. 6 - p. 467

897. THICKNESS OF THE EPIDERMIS:
A. Palms
B. Soles Ref. 6 - p. 481

898. NUMBER OF CORNIFIED CELLS:
A. Stratum granulosum
B. Stratum lucidum Ref. 6 - p. 483

899. PRESENCE OF SEBACEOUS GLANDS:
A. Palms
B. Soles Ref. 6 - p. 497

900. ABUNDANCE OF MUCOUS CELLS:
A. Sublingual gland
B. Parotid gland Ref. 6 - p. 521

901. DEGREE OF HARDNESS:
A. Enamel
B. Bone Ref. 6 - p. 532

902. CONTRIBUTORS TO THE SECRETION OF GASTRIC JUICE:
A. Gastric glands
B. Brunner's glands Ref. 6 - p. 549

903. DIGESTIVE ENZYMES LOCALIZED IN THE PANETH CELLS:
A. Trypsin
B. Pepsinogen Ref. 6 - p. 565

904. SIMILARITY TO PERICYTES OF CAPILLARIES:
A. Deep cells or mesangial cells
B. Podocytes Ref. 6 - pp. 654,656

905. NUMBER OF AUTOSOMES:
A. Human male
B. Human female Ref. 6 - p. 695

906. LOCATION OF OVARIAN FOLLICLES:
A. Cortex
B. Medulla
Ref. 6 - p. 728

907. THICKNESS OF TUNICA IN WALLS OF ARTERIOLES:
A. Tunica media
B. Tunica adventitia
Ref. 6 - p. 367

908. FOUND IN CONCAVITIES IN THE SURFACE OF BONE (HOWSHIP'S LACUNAE):
A. Osteoblasts
B. Osteoclasts
Ref. 6 - p. 233

909. DENSENESS OF CONNECTIVE TISSUE:
A. Dura mater
B. Pia mater
Ref. 6 - p. 347

910. ELEMENTS OF ARACHNOID AND PIA MATER:
A. Collagenous fibers
B. Reticular fibers
Ref. 6 - p. 347

SELECT THE ITEM FROM COLUMN I WHICH IS UNRELATED TO THE FOUR OTHER ITEMS IN THIS COLUMN, CIRCLE THIS LETTER. IDENTIFY THE ITEM IN COLUMN II TO WHICH THE FOUR ITEMS IN COLUMN I ARE RELATED, CIRCLE THIS NUMBER.

	COLUMN I	COLUMN II
911.	A. Sex chromatin B. Chromosomes C. Nucleoli D. Nuclear membrane E. Mitochondria	1. Constituents of cell nucleus 2. Phases of amitosis 3. Constituents of cytoplasm Ref. 6 - pp. 53-59
912.	A. Amitosis B. Prophase C. Spindle D. Centriole E. Anaphase	1. Mitosis 2. Budding 3. Binary fission Ref. 6 - pp. 60-72
913.	A. Mouth B. Anal mucous membrane C. Cornea D. Vagina E. A part of the epiglottis	1. Pseudostratified columnar epithelium 2. Stratified squamous epithelium 3. Simple columnar ciliated epithelium Ref. 6 - p. 82
914.	A. Howell-Jolly bodies B. Rings of Cabot C. Azurophilic granules D. Granules of Ehrlich-Heinz E. Romanowsky stain	1. Erythrocytes 2. Monocytes 3. Mast cells Ref. 6 - pp. 112-116

	COLUMN I	COLUMN II
915.	A. Fibroblasts B. Histiocytes C. Eosinophils D. Osteocytes E. Mast cells	1. Dense connective tissue 2. Bone 3. Loose connective tissue Ref. 6 - pp. 144-151
916.	A. Periosteum B. Isogenous cell groups C. Collagenous fibers D. Chondroitin sulfate E. Absence of blood vessels	1. Tendons 2. Hyaline cartilage 3. Bone Ref. 6 - pp. 212-215
917.	A. May reach length of 0. 5 mm. B. Distinct sarcolemma C. Eccentrically placed nucleus D. Myofibrils E. Sarcoplasm	1. Cardiac muscle cells 2. Smooth muscle cells 3. Striated muscle cells Ref. 6 - pp. 263-270
918.	A. Lines of Retzius B. Collagenous fibrils C. Lines of Owen D. "Interglobular" spaces E. Protoplasmic fibers of Tomes	1. Cementum 2. Enamel 3. Dentin Ref. 6 - pp. 530-532
919.	A. Zymogenic cells B. Oxyphile cells C. Parietal cells D. Mucous neck cells E. Argentaffin cells	1. Esophagus 2. Stomach 3. Small intestine Ref. 6 - p. 550
920.	A. Low columnar cells B. Cells of Paneth C. Argentaffin cells D. Basket cells E. Mitosis	1. Crypts of Lieberkühn 2. Gastric glands 3. Glands of Brunner Ref. 6 - p. 564
921.	A. Columnar cells with striated border B. Goblet cells C. Oxyphile cells D. Argentaffin cells E. Valves of Kerking	1. Small intestine 2. Stomach 3. Esophagus Ref. 6 - pp. 560-567

HISTOLOGY

	COLUMN I	COLUMN II
922.	A. Intercalated vein	1. Spleen
	B. Collecting vein	2. Liver
	C. Central vein	3. Pancreas
	D. Sinusoid	
	E. Red pulp	Ref. 6 - p. 588
923.	A. Medullary rays (of Ferrein)	1. Adrenal gland
	B. Cortex	2. Kidney
	C. Medulla	3. Male urethra
	D. Area cribrosa	
	E. Epinephrine	Ref. 6 - pp. 652-664
924.	A. Axis cylinder	1. Unmyelinated nerve fibers
	B. Myelin	2. All nerve fibers
	C. Schwann's cells	3. Myelinated nerve fibers
	D. Nodes of Ranvier	
	E. Neuroglia	Ref. 6 - pp. 319-324
925.	A. Pyramidal cells	1. Nerve cells
	B. Ependyma	2. Connective tissue
	C. Satellite cells	3. "Neuroglia"
	D. Cells of Schwann	
	E. Microglia	Ref. 6 - pp. 335-338
926.	A. Aorta	1. Elastic type artery
	B. Brachial	2. Muscular type artery
	C. Innominate	3. Arteriole
	D. Subclavian	
	E. Pulmonary	Ref. 6 - p. 366
927.	A. Collagenous bundles	1. Parathyroid gland
	B. Sinusoids	2. Pars distalis of hypophysis
	C. Paneth cells	3. Gastric glands
	D. Chief cells	
	E. Chromophile cells	Ref. 6 - pp. 430-433
928.	A. Oxyphile cells	1. Thyroid gland
	B. Principal cells	2. Pineal gland
	C. Distinct basement membrane	3. Parathyroid gland
	D. Follicles with colloid	
	E. "Water-clear" cells	Ref. 6 - pp. 454-457

COLUMN I	COLUMN II

929. A. Huxley's layer
B. Outermost layer of Henle
C. Glassy membrane
D. Hyponychium
E. Cortex

1. Hair
2. Nails
3. Kidney

Ref. 6 - pp. 491-494

930. A. 2 to 3 mm. long
B. Connective tissue core
C. No true keratin
D. Hard scales
E. Glands of von Ebner

1. Circumvallate papillae
2. Fungiform papillae
3. Filiform papillae

Ref. 6 - p. 513

931. A. Sharpey's fibers
B. 18 to 20 cm. long
C. Pars membranacea urethra
D. Lacunae of Morgagni
E. Glands of Littre

1. Male urethra
2. Female urethra
3. Ureter

Ref. 6 - pp. 678-680

932. A. Vascular cones
B. Hydatid Morgagni
C. Alternating groups of tall and low cells
D. Thin basement membrane
E. Circularly arranged smooth muscle

1. Rete testis
2. Ductuli efferentes
3. Ductus epididymis

Ref. 6 - pp. 709-710

933. A. Membrana granulosa
B. Glassy membrane
C. About 45 microns in diameter
D. Theca of the follicle
E. Corona radiata

1. Primary follicles
2. Growing follicles
3. Mature Graafian follicle

Ref. 6 - pp. 731-732

934. A. Glands usually absent
B. Stratified squamous epithelium
C. Striated muscle fibers
D. No true muscular coat
E. Lymph nodules

1. Uterus
2. Oviduct
3. Vagina

Ref. 6 - pp. 763-764

935. A. Capsule of Tenon
B. Membrane of Descemet
C. Membrane of Bowman
D. Melanocytes
E. Afferent arteriole

1. Kidney
2. Eye
3. Ear

Ref. 6 - pp. 776-781

IN THE FOLLOWING QUESTIONS ARRANGE THE ITEMS IN PROPER SEQUENCE:

936. VARIOUS STAGES OF MITOSIS:
A. Prometaphase
B. Interphase
C. Telophase
D. Advancing prophase
E. Anaphase Ref. 6 - p. 62

937. STAGES IN DEVELOPMENT OF AN ERYTHROCYTE:
A. Hemocytoblast
B. Basophilic erythroblast
C. Polychromatophilic erythroblast
D. Normoblast
E. Normoblast with pyknotic nucleus Ref. 6 - p. 186

938. REGIONS OF ADRENAL GLAND, FROM OUTERMOST TO INNERMOST AREA:
A. Zona fasciculata
B. Zona glomerulosa
C. Capsule
D. Zona reticularis
E. Medulla Ref. 6 - p. 461

939. LAYERS OF A HAIR, FROM OUTERMOST TO INNERMOST LAYER:
A. Cuticle
B. Henle's layer
C. External root sheath
D. Huxley's layer
E. Glassy membrane Ref. 6 - p. 500

940. PATHWAY TAKEN BY A SPERM FROM SITE OF FORMATION:
A. Rete testis
B. Seminiferous tubule B A D C E
C. Ductus epididymis
D. Canaliculi efferentes
E. Ductus deferens Ref. 6 - p. 716

941. ENCOUNTERED IN SECTION OF INTESTINAL TRACT, FROM LUMEN OUTWARD:
A. Meissner's plexus
B. Lamina propria
C. Muscularis mucosa
D. Glands of Lieberkühn
E. Myenteric plexus of Auerbach Ref. 6 - p. 561

942. PARTS OF A "TYPICAL" NEURON; FOLLOWING PATHWAY OF IMPULSE TRANSMISSION:
A. Telodendrons
B. Perikaryon
C. Axon hillock
D. Node of Ranvier
E. Dendrites Ref. 6 - p. 305

943. STAGES OF ENDOCHONDRIAL BONE FORMATION; AS SEEN IN MICROSCOPIC SECTION:
A. Proliferating cartilage cells
B. Hypertrophic cartilage cells
C. Provisional calcification
D. Primary spongiosa
E. Secondary spongiosa Ref. 6 - p. 238

944. LAYERS OF THE ESOPHAGUS FROM LUMEN OUTWARD:
A. Inner longitudinal muscle
B. Lamina propria
C. Circular muscle
D. Lamina muscularis mucosae
E. Submucosa Ref. 6 - p. 545

945. SECTION OF HUMAN CORNEA STARTING FROM ENDOTHELIUM:
A. Bowman's membrane
B. Corneal epithelium
C. Endothelium
D. Descemet's membrane
E. Stroma Ref. 6 - p. 780

FOLLOWING EACH GROUP OF FIVE CHOICES THERE ARE EITHER FIVE STATEMENTS OR FIVE ITEMS. MATCH THE STATEMENTS OR ITEMS WITH THE MOST APPROPRIATE CHOICE GIVEN. EACH CHOICE MAY BE USED ONLY ONCE.

A. Periodic acid-Schiff reaction
B. Feulgen reaction
C. Gomori-Takamatsu procedure
D. Toluidine blue
E. Sudan black and Nile blue

946. ___ Owes its specificity to the presence of an aldehyde in deoxyribose

947. ___ Used for identification of carbohydrates

948. ___ For identification of alkaline phosphotase

949. ___ Used to demonstrate basophilic components of the cell

950. ___ Frequently used to demonstrate fatty components of the cell
Ref. 6 - pp. 11, 10-17

A. Phase contrast microscope
B. Interference microscope
C. Polarizing microscope
D. Fluorescence microscope
E. Electron microscope

951. ___ Utilizes principle of sending two separate light beams through the specimen. Can be used for contour mapping of small objects

952. ___ Indirectly provides information about structural arrangement at the molecular level; applied to living cells

953. ___ Shaped magnetic fields or electric fields replace glass lenses

954. ___ Renders visible certain components of living cells that usually require fixation and staining

955. ___ Light of one wavelength is used to illuminate the specimen. The image is due to emission from molecules that have absorbed the primary exciting light and then re-emitted light of longer wavelength and lower energy.

Ref. 6 - pp. 22-25

A. Plasmalemma
B. Ergastoplasm
C. Mitochondria
D. Golgi apparatus
E. Centrosome

956. ___ Basophilic material in cytoplasm, chromidial substance, ribonucleoprotein, endoplasmic reticulum

957. ___ Internal reticular apparatus, juxtanuclear position, consists of curved parallel arrays of flattened saccules

958. ___ Discovered by Altmann (1890), oxidative phosphorylation, enzymes of Krebs citric acid cycle in matrix

959. ___ Term introduced by Boveri (1888), associated with cell division

960. ___ Selective permeability, bimolecular layer of mixed lipids between two layers of adsorbed protein

Ref. 6 - pp. 40-49

A. Simple squamous epithelium
B. Simple columnar epithelium
C. Stratified squamous epithelium
D. Pseudostratified columnar epithelium
E. Simple cuboidal epithelium

961. ___ Found in descending limb of loop of Henle

962. ___ Found in thyroid, free surface of ovary

963. ___ Common in excretory ducts of many glands

964. ___ Found in a part of the epiglottis, cornea, vagina

965. ___ Found in male urethra and large excretory duct of parotid

Ref. 6 - pp. 80-83

A. Small lymphocyte
B. Monocyte
C. Eosinophile
D. Basophile
E. Neutrophile

966. ___ Relatively large nucleus with scant cytoplasm

967. ___ Relatively small nucleus with abundant cytoplasm

968. ___ Nucleus usually has two oval lobes

969. ___ Difficult to find in human blood

970. ___ Most important leukocyte in vertebrates

Ref. 6 - pp. 117-123

IDENTIFY COMPONENTS OF THE "TYPICAL CELL" AND SOME OF THE STAGES OF MITOSIS.

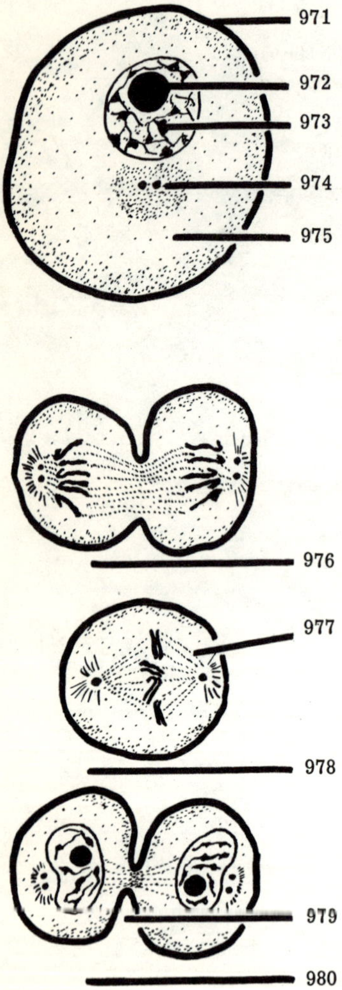

A. Chromatin
B. Cell membrane
C. Spindle
D. Metaphase
E. Nucleolus
F. Telophase
G. Mitochondria
H. Cytoplasm
I. Cytokinesis
J. Prophase
K. Anaphase
L. Golgi apparatus
M. Centrioles
N. Astral rays
O. Equitorial plane

Ref. 6 - pp. 59,62

IDENTIFY THE COMPONENTS OF THE SCHEMATIC-COMPOSITE OF THE OVARY.

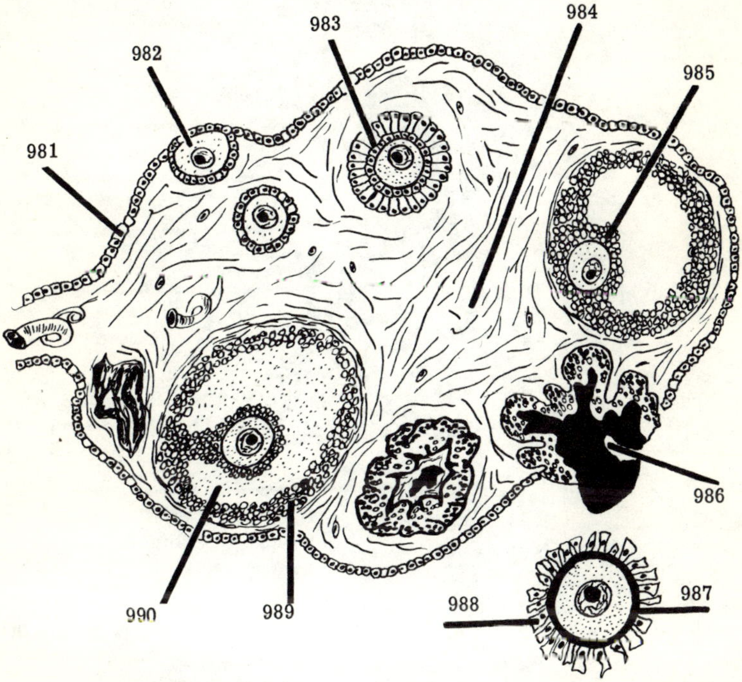

A. Primordial follicle
B. Corpus luteum
C. Corpus albicans
D. Theca folliculi
E. Medulla
F. Follicular epithelium
G. Cumulus oöphorus
H. Follicular liquid
I. Membrana granulosa
J. Theca externa
K. Growing follicle
L. Corpus atreticum
M. Zona pellucida
N. Cortex
O. Germinal epithelium

Ref. 5 - pp. 585-589

IDENTIFY THE STRUCTURES THROUGH THE SECTION OF A DEVELOPING LOWER INCISOR.

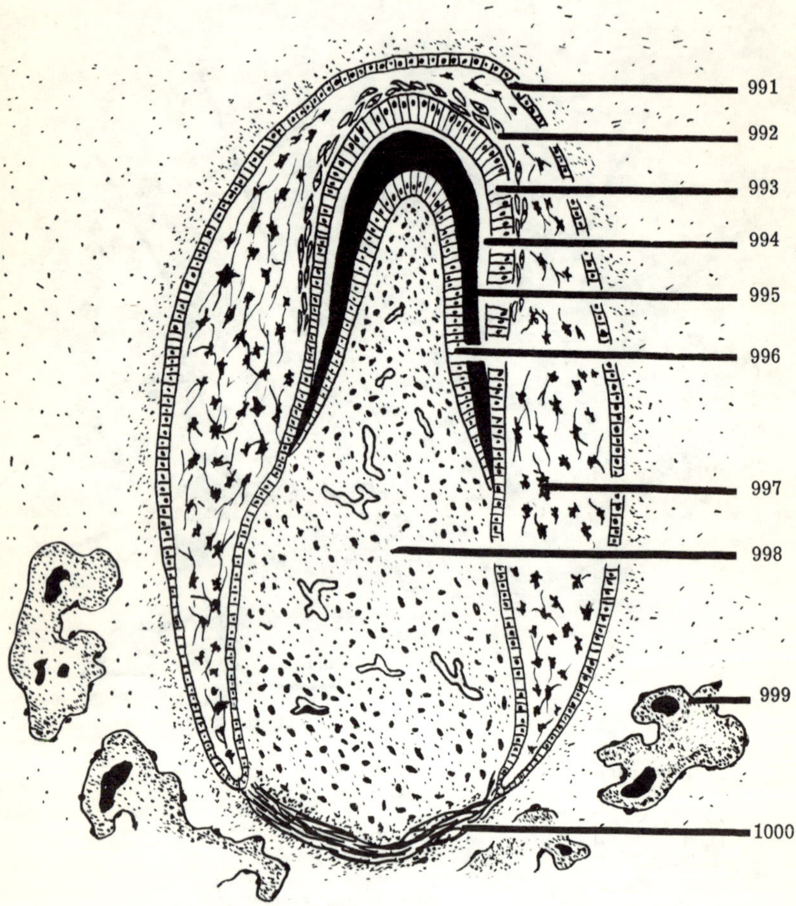

A. Internal enamel epithelium
B. External enamel epithelium
C. Dentin
D. Primordium of odontoblasts
E. Gingiva
F. Enamel pulp
G. Dental follicle
H. Epithelial attachment of Gottlieb
I. Dental pulp
J. Cementum
K. Chondrocytes
L. Enamel cap
M. Hertwig's sheath
N. Alveolar bone
O. Stratum intermedium

Ref. 6 - pp. 530, 540

IDENTIFY THE COMPONENTS OF THE HUMAN COCHLEAR DUCT, AS INDICATED BY THE GUIDE LINES.

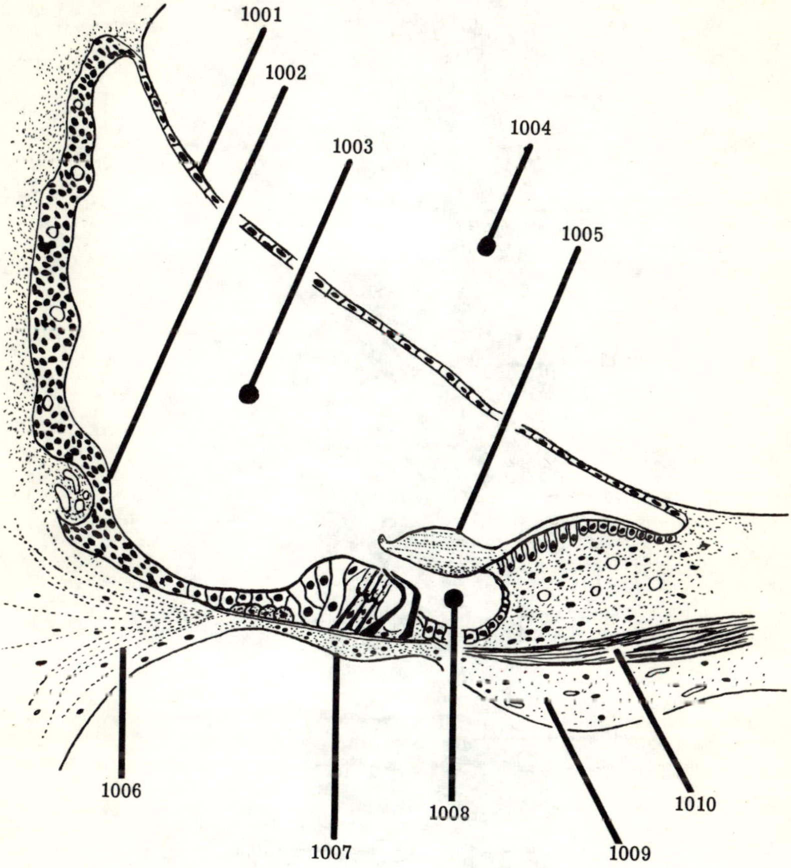

A. Stria vascularis
B. Cochlear nerve
C. Vestibular membrane
D. Tympanic lip
E. Scala vestibuli
F. Cochlear duct
G. Vestibular lip
H. External spiral sulcus
I. Habenula perforata
J. Tectorial membrane
K. Internal spiral tunnel
L. Basilar membrane
M. Spiral ligament
N. Hensen's stripe
O. Limbus lamina spiralis

Ref. 6 - pp. 823,825-827

IDENTIFY THE STRUCTURES INDICATED BY THE GUIDE LINES IN THE SCHEMATIC SECTION (LONGITUDINAL) OF THE HUMAN DUODENUM.

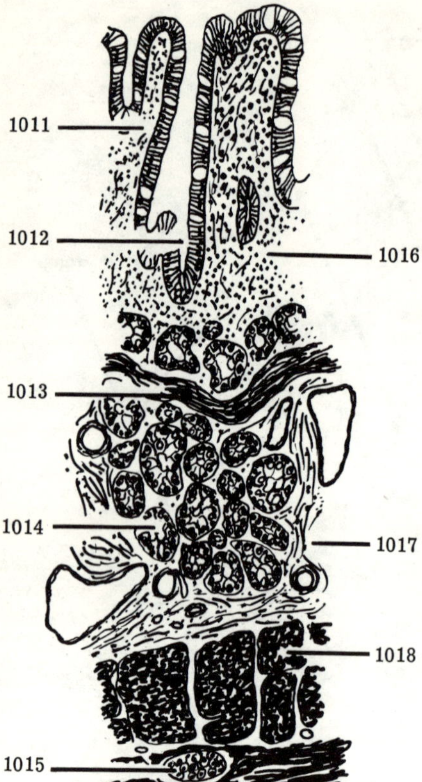

A. Submucosa
B. Longitudinal muscle
C. Villus
D. Muscularis mucosae
E. Endothelium
F. Adventitia
G. Mesothelium of serosa
H. Crypt of Lieberkühn
I. Circular muscle
J. Brunner's glands
K. Parietal layer
L. Plexus mysentericus
M. Lamina propria
N. Myomysium
O. Sharpey's fibers

Ref. 6 - pp. 561, 563

SECTION I - EMBRYOLOGY, HISTOLOGY, GROSS ANATOMY

FOR EACH OF THE FOLLOWING MULTIPLE CHOICE QUESTIONS SELECT THE ONE MOST APPROPRIATE ANSWER:

1021. THE TERMS LEPTOMENINX OR LEPTOMENINGES ARE USED TO DESIGNATE:
A. Pia mater and arachnoid
B. Pia mater and dura mater
C. Dura mater and arachnoid
D. Pia mater, dura mater and arachnoid
E. Dura mater and cranial bone periosteum
Ref. 8 - p. 12

1022. THE SPINAL CORD IS ANCHORED TO THE DURA MATER BY THE:
A. Filum terminale
B. Pia mater
C. Denticulate ligaments
D. Conus medularis
E. Galea aponeurotica Ref. 8 - p. 15

1023. THE SPINAL CORD EXTENDS FROM FORAMEN MAGNUM TO THE LOWER BORDER OF THE:
A. Tenth thoracic vertebra
B. Eleventh thoracic vertebra
C. First lumbar vertebra
D. Third lumbar vertebra
E. Fifth lumbar vertebra Ref. 8 - p. 26

1024. THERE ARE ___________ PAIRS OF NERVES WHICH ARISE FROM THE SPINAL CORD:
A. 25
B. 27
C. 31
D. 33
E. 35 Ref. 8 - p. 27

1025. THE FLOOR OF THE FOURTH VENTRICLE IS FORMED BY THE POSTERIOR SURFACE OF THE MEDULLA AND THE:
A. Superior medullary velum
B. Inferior medullary velum
C. Brachium pontis
D. Calamus scriptorius
E. Pontine tegmentum Ref. 8 - p. 34

1026. THE LATERAL SULCUS OF THE CEREBRUM STARTS ON THE INFERIOR SURFACE AS A DEEP CLEFT, THE:
A. Central sulcus of Rolando
B. Sylvian fossa
C. Transverse fissure
D. Fossa of Monro
E. Fossa of Magendie Ref. 8 - p. 44

SECTION I - EMBRYOLOGY, HISTOLOGY, GROSS ANATOMY

1027. BURIED IN THE LATERAL SULCUS OF THE CEREBRAL HEMISPHERE IS THE:
A. Occipital lobe
B. Splenium
C. Temporal gyri of Heschl
D. Insula (Island of Reil)
E. Pars triangularis — Ref. 8 - p. 46

1028. THE TRIANGULAR AREA BETWEEN FORNIX AND CORPUS CALLOSUM IS OCCUPIED BY THE:
A. Insula
B. Septum pellucidum
C. Genu
D. Uncus
E. Lateral geniculate body — Ref. 8 - p. 48

1029. THE MIDDLE CEREBRAL ARTERY IS A CONTINUATION OF THE:
A. External carotid
B. Internal carotid
C. Vertebral
D. Anterior temporal
E. Basilar — Ref. 8 - p. 67

1030. THE MEDULLA AND PONS ARE SUPPLIED BY ALL OF THE FOLLOWING ARTERIES, EXCEPT FOR THE:
A. Anterior spinal
B. Posterior spinal
C. Branches of basilar
D. Branches of vertebral
E. Middle cerebral — Ref. 8 - p. 74

1031. THE SPINAL CORD IS APPROXIMATELY ____________ CM. IN LENGTH:
A. 25
B. 35
C. 45
D. 55
E. 65 — Ref. 7 - p. 22

1032. THE FILUM TERMINALE IS COMPOSED CHIEFLY OF:
A. Pia mater
B. Dura mater
C. Cerebrospinal fluid
D. White matter
E. Ganglion cells — Ref. 7 - p. 24

1033. THE ROOTS OF THE SPINAL NERVES BELOW THE FIRST LUMBAR IS CALLED THE:
A. Filum terminale
B. Conus medullaris
C. Cauda equina
D. Lumbar plexus
E. Sacral plexus — Ref. 7 - p. 27

SECTION I - EMBRYOLOGY, HISTOLOGY, GROSS ANATOMY

1034. WHICH OF THE FOLLOWING BONES DOES NOT JOIN THE OTHERS TO FORM THE PTERION?:
A. Frontal
B. Temporal
C. Sphenoid
D. Parietal
E. Occipital Ref. 7 - p. 29

1035. THE CEREBELLUM AND CEREBRAL HEMISPHERES ARE SEPARATED BY THE:
A. Falx cerebri
B. Tentorium cerebelli
C. Galea aponeurotica
D. Great vein of Galen
E. Pons Ref. 7 - p. 30

1036. THE MAIN CHANNELS OF COMMUNICATION BETWEEN THE SPINAL CORD AND MEDULLA OBLONGATA WITH THE CEREBELLUM ARE THE:
A. Cerebral peduncles
B. Superior olives
C. Inferior olives
D. Restiform bodies
E. Pontine fibers Ref. 7 - p. 33

1037. THE CENTRAL CANAL OF THE SPINAL CORD OPENS OUT INTO THE:
A. Third ventricle
B. Fourth ventricle
C. Cisterna magna
D. Aqueduct of Sylvius
E. Foramen of Magendie Ref. 7 - p. 34

1038. THE ANTERIOR MEDIAN FISSURE OF THE SPINAL CORD ENDS IN A PIT KNOWN AS THE:
A. Foramen caecum
B. Posterior lateral sulcus
C. Foramen of Magendie
D. Aqueduct of Sylvius
E. Lateral ventricle Ref. 7 - p. 35

1039. THE SUPERIOR CEREBELLAR PEDUNCLES CONNECT THE CEREBELLUM WITH THE:
A. Restiform body
B. Inferior olive
C. Red nucleus
D. Portio major
E. Anterior medullary velum Ref. 7 - p. 38

SECTION I - EMBRYOLOGY, HISTOLOGY, GROSS ANATOMY

1040. WHICH OF THE FOLLOWING DOES NOT ENTER INTO THE FORMATION OF THE LATERAL BOUNDARIES OF THE RHOMBOID FOSSA?:
A. Brachia conjunctiva
B. Restiform bodies
C. Cuneate tubercles
D. Clavae
E. Anterior medullary velum Ref. 7 - p. 39

1041. BETWEEN THE FUNICULUS SEPARANS AND THE CLAVA IS THE:
A. Tela chorioidea
B. Area postrema
C. Facial colliculus
D. Calamus scriptorus
E. Sulcus limitans Ref. 7 - p. 41

1042. WHICH OF THE FOLLOWING BELONG TO THE PALEOCEREBELLUM?:
A. Pyramis
B. Folium
C. Tuber
D. All of these
E. None of these Ref. 7 - p. 45

1043. FOLDS OF THE CEREBRAL CORTEX FOLDING IN OVER THE INSULA ARE CALLED THE:
A. Temporal lobe
B. Corpus callosum
C. Opercula
D. Commissure of the fornix
E. Pallium Ref. 7 - p. 53

1044. CELLS OF THE SYMPATHETIC GANGLIA ARE DERIVED FROM THE:
A. Neural crest
B. Spinal cord
C. Notochord
D. Oligodendrocytes
E. Neural ependyma Ref. 7 - p. 95

1045. NERVE FIBERS OF THE SPINAL CORD ARE DEVOID OF:
A. Axons
B. Neurofibrils
C. Dendrites
D. Telodendrons
E. Neurilemma sheaths Ref. 7 - p. 119

1046. WHICH OF THE FOLLOWING TERMS DOES NOT APPLY TO A BUNDLE OF NERVE FIBERS?:
A. Peduncle
B. Gray matter
C. Funiculi
D. Path
E. Fasciculi Ref. 7 - p. 20

SECTION I - EMBRYOLOGY, HISTOLOGY, GROSS ANATOMY

1047. GROSSLY THE SPINAL CORD PRESENTS TWO SWELLINGS:
A. Cervical and thoracic
B. Cervical and lumbar
C. Thoracic and lumbar
D. Thoracic and sacral
E. Lumbar and sacral — Ref. 7 - p. 23

1048. THE RHOMBENCEPHALON IS MADE UP OF THE:
A. Cerebrum, cerebellum, pons
B. Cerebellum, pons, spinal cord
C. Spinal cord, medulla oblongata, cerebellum
D. Medulla oblongata, cerebellum, pons
E. None of the above — Ref. 7 - p. 30

1049. SURROUNDING THE FOURTH VENTRICLE ARE THE:
A. Medulla oblongata, pons, cerebellum
B. Cerebellum, pons, thalamus
C. Pons, cerebellum, cerebrum
D. Cerebellum, medulla oblongata, spinal cord
E. None of the above — Ref. 7 - p. 33

1050. ORIGINATING AT THE CEREBRAL AQUEDUCT AND RUNNING THROUGH BOTH SUPERIOR AND INFERIOR FOVAE IS THE:
A. Foramen of Magendie
B. Foramen of Luschka
C. Sulcus limitans
D. Clava
E. Nucleus intercalatus — Ref. 7 - p. 40

1051. THE ARTERIAL CIRCLE OF WILLIS IS FORMED BY BRANCHES OF THE ____________ AND INTERNAL CAROTID ARTERIES:
A. External carotid
B. Vertebral
C. Basilar
D. Middle meningeal
E. Anterior temporal — Ref. 7 - p. 80

1052. WHICH OF THE FOLLOWING CELL LINES ARE NOT DERIVED FROM MEDULLARY EPITHELIUM:
A. Ependyma
B. Astrocytes
C. Oligodendrocytes
D. Microglia
E. Neurons — Ref. 7 - p. 93

1053. MULTIPOLAR NEURONS USUALLY CONTAIN ______________ AXONS:
A. 1
B. 2
C. 3
D. 4
E. 5 — Ref. 7 - p. 100

SECTION I - EMBRYOLOGY, HISTOLOGY, GROSS ANATOMY

1054. CELLS WITH LONG AXONS ARE CLASSIFIED AS __________ NEURONS:
A. Unipolar
B. Bipolar
C. Multipolar
D. Golgi type I
E. Golgi type II Ref. 7 - p. 101

1055. BRANCHES OF AN AXON MAY END ON THE SURFACE OF THE PERIKARYON OR DENDRITES OF ANOTHER NEURON IN THE FORM OF:
A. Gliosomes
B. Motor end plates
C. Boutons terminaux
D. Foot plates
E. None of the above Ref. 7 - p. 109

1056. MYELINATED FIBERS OF THE BRAIN DIFFER FROM THOSE OF THE PERIPHERAL NERVES IN THAT THE FORMER DO NOT POSSESS:
A. Axons
B. Neurilemma sheaths
C. Neuroglia fibers
D. Nodes of Ranvier
E. Perikaryons Ref. 7 - p. 114

1057. THE SUPERIOR SAGITTAL SINUS EXTENDS FROM THE INTERNAL OCCIPITAL PROTUBERANCE TO:
A. Foramen cecum
B. Foramen of Magendi
C. Foramen rotundum
D. Foramen of Winslow
E. None of the above Ref. 8 - p. 81

1058. THE CAVERNOUS SINUS IS LOCATED ON THE SIDE OF THE:
A. Parietal bone
B. Temporal bone
C. Sphenoid bone
D. Occipital bone
E. Frontal bone Ref. 8 - p. 82

1059. THE OPTIC NERVES ARE ASSOCIATED WITH THE:
A. Telencephalon
B. Diencephalon
C. Metancephalon
D. Mesencephalon
E. Myelencephalon Ref. 8 - p. 102

1060. NERVE TISSUE HAS A STRONG AFFINITY FOR:
A. Strong zinc solutions
B. Weak gold solutions
C. Weak silver solutions
D. Strong osmium solutions
E. Weak lead solutions Ref. 8 - p. 108

SECTION I - EMBRYOLOGY, HISTOLOGY, GROSS ANATOMY

1061. ALL OF THE FOLLOWING ARE ASSOCIATED WITH BIPOLAR NEURONS, EXCEPT FOR:
A. Olfactory receptor endings
B. Auditory receptor endings
C. Vestibular receptor endings
D. Visual receptor endings
E. Touch receptor endings Ref. 8 - p. 114

1062. NISSL BODIES ARE CHARACTERIZED BY:
A. Appearance as deeply staining granules
B. Presence in cell bodies and dendrites
C. Absence in axon and axon hillock
D. All of the above
E. None of the above Ref. 8 - p. 121

1063. NEUROFIBRILS ARE FOUND IN:
A. Unipolar neurons
B. Bipolar neurons
C. Multipolar neurons
D. All of the above
E. None of the above Ref. 8 - p. 121

1064. "A" FIBERS ARE CHARACTERIZED BY ALL OF THE FOLLOWING, EXCEPT:
A. Unmyelinated
B. Range in diameter from 1 to 20 micra
C. Conduct at rates of 5 to 120 meters/second
D. Some transmit motor impulses to skeletal muscles
E. Some transmit afferent stretch muscles
Ref. 8 - p. 135

1065. IN ITS BROADEST SENSE, ALL OF THE FOLLOWING ARE NEUROGLIA, EXCEPT FOR:
A. Satellite cells
B. Purkinje cells
C. Protoplasmic astrocytes
D. Ependyma
E. Oligodendrocytes Ref. 8 - p. 149

SECTION I - EMBRYOLOGY, HISTOLOGY, GROSS ANATOMY

ANSWER THE FOLLOWING QUESTIONS ACCORDING TO THE KEY OUTLINED BELOW:

A. If both statement and reason are true and related cause and effect
B. If both statement and reason are true but not related cause and effect
C. If the statement is true but the reason is false
D. If the statement is false and the reason is true
E. If both statement and reason are false

1066. The retina is genetically part of the brain

because

It develops as an outpocketing from the walls of the telencephalon.
Ref. 7 - p. 13

1067. The third ventricle is reduced to a small vertical cleft

because

Of the great growth of the thalamus. Ref. 7 - p. 16

1068. The medulla oblongata is not considered as a derivative of the primary vesicles of the brain

because

It is derived from the developing spinal cord.
Ref. 7 - p. 17

1069. In performing a lumbar puncture ("spinal tap") the needle must enter the subdural space

because

Cerebrospinal fluid is normally found in this space.
Ref. 7 - p. 23

1070. Most of the cervical roots run almost laterally

because

The cervical part of the spinal cord changes its position but very little as development progresses from the fetal stage.
Ref. 7 - pp. 26-27

1071. The cerebral hemispheres form the largest part of the prosencephalon

because

It occupies the anterior and middle cranial fossae.
Ref. 7 - pp. 30-31

1072. Brain weight is an important factor in determining intelligence

because

The brain weighs less in older persons. Ref. 7 - p. 34

1073. The diamond-shaped cavity of the rhombencephalon is known as the third ventricle

because

It is situated in the third portion of the medulla oblongata.
Ref. 7 - p. 38

SECTION I - EMBRYOLOGY, HISTOLOGY, GROSS ANATOMY

1074. The fourth ventricle communicates with the subarachnoid space through the lateral aperture, foramen of Magendie

because

Through this aperature cerebrospinal fluid may enter the subarachnoid space. Ref. 7 - p. 39

1075. The optic nerve fibers are devoid of neurilemma sheaths

because

This is the only pair of nerves associated with the diencephalon.
Ref. 7 - p. 48

1076. The cerebral hemispheres are incompletely separated from each other

because

At the bottom of the longitudinal fissure they are connected by the corpus callosum. Ref. 7 - p. 54

1077. Within the cranium the dura mater consists of two layers

because

The dura and periosteum are fused in this area.
Ref. 7 - p. 71

1078. The cerebrospinal fluid is of low specific gravity

because

It is in osmotic blance with the blood. Ref. 7 - p. 76

1079. The two vertebral arteries are a major source of blood for the brain

because

They enter the skull through foramen rotundum.
Ref. 7 - p. 77

1080. Oligodendrocytes are the smallest of the macroglial cells

because

They are found in both the gray and white matter.
Ref. 8 - p. 149

1081. Astrocytes are found only in white matter

because

They are involved as scavengers of the nervous system.
Ref. 8 - pp. 152,154

1082. The subarachnoid space is in direct communication with the third ventricle

because

The foramen of Magendie opens into it.
Ref. 8 - p. 18

1083. The mesencephalon is the smallest and least differentiated of the brain divisions

because

The midbrain contains only connections to the medulla oblongata.
Ref. 8 - pp. 38-39

SECTION I - EMBRYOLOGY, HISTOLOGY, GROSS ANATOMY

1084. The thalamus forms the greater part of the diencephalon

because

The hypothalamus forms the lateral wall and floor of the third ventricle.

Ref. 8 - pp. 40-41

1085. Complete thrombosis of the basilar artery is generally fatal

because

It causes massive frontal lobe damage.

Ref. 8 - p. 75

SECTION I - EMBRYOLOGY, HISTOLOGY, GROSS ANATOMY

IDENTIFY THE STRUCTURES IN A CORONAL SECTION THROUGH THE SUPERIOR SAGITTAL SINUS, AS INDICATED BY THE GUIDE LINES.

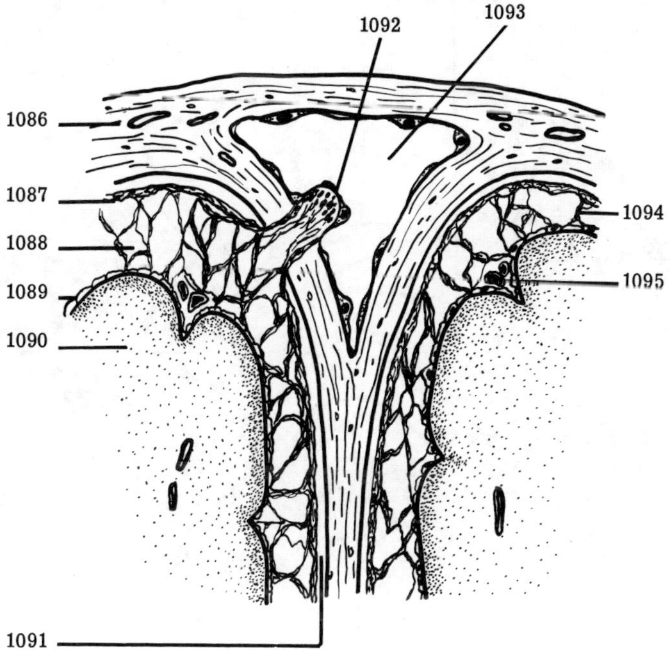

A. Subarachnoid space
B. Sharpey's fibers
C. Falx cerebri
D. Tentorium cerebelli
E. Pia mater
F. Periosteum
G. Cerebral vein
H. Subdural space
I. Arachnoid villus
J. Arachnoid membrane
K. Cerebral cortex
L. Arachnoid trabeculae
M. Epidural space
N. Superior sagittal sinus
O. Dura mater

Ref. 7 - pp. 72-73

SECTION I - EMBRYOLOGY, HISTOLOGY, GROSS ANATOMY

IDENTIFY THE STRUCTURES SEEN IN A MEDIAL SAGITTAL SECTION OF THE BRAIN.

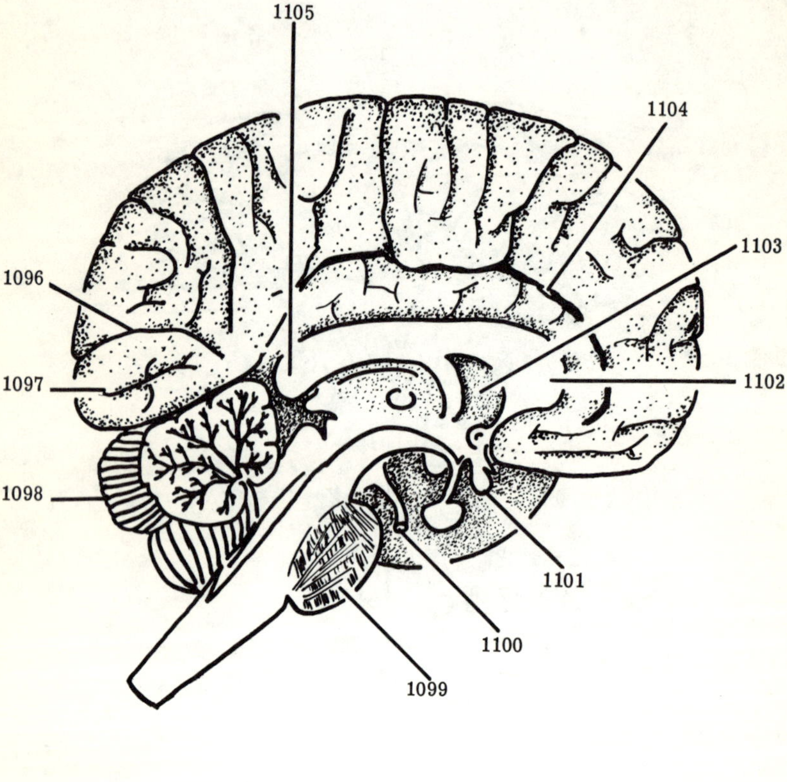

A. Calcarine fissure
B. Optic chiasm
C. Sulcus cinguli
D. Cerebellum
E. Splenium of corpus callosum
F. Oculomotor nerve
G. Mammillary body
H. Septum pellucidum
I. Occipital pole
J. Pons
K. Genu of corpus callosum
L. Trigeminal nerve
M. Pineal body
N. Thalamus
O. Body of fornix
P. Frontal lobe

Ref. 7 - p. 31

SECTION I - EMBRYOLOGY, HISTOLOGY, GROSS ANATOMY

IDENTIFY THE PARTS OF A "TYPICAL NEURON".

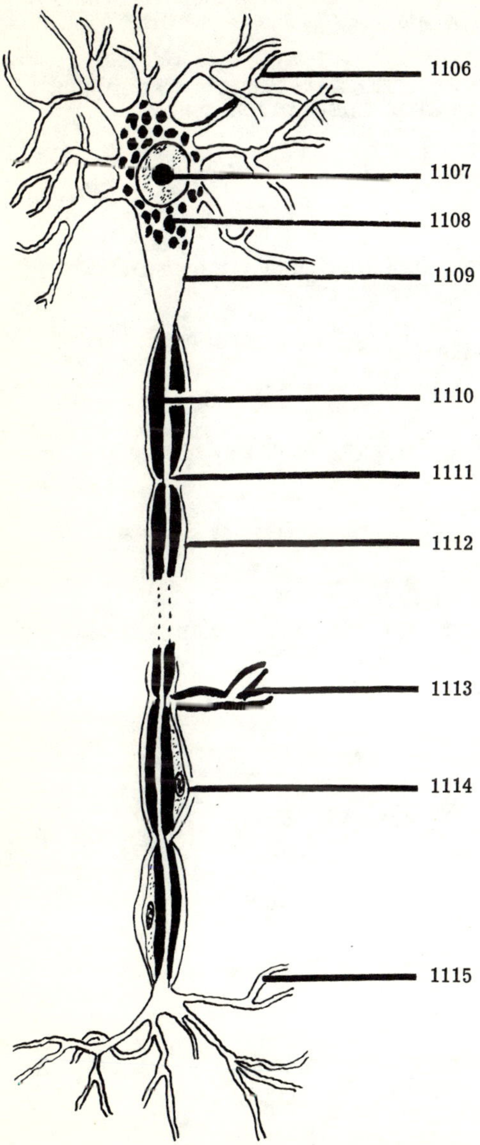

A. Nucleolus
B. Neurilemma
C. Nucleus
D. Axon hillock
E. Telodendron
F. Dendrite
G. Schwann cell
H. Nissl substance
I. Node of Ranvier
J. Myelin
K. Golgi apparatus
L. Collateral branch
M. Myoneurium
N. Epineurium
O. Gliosome

Ref. 7 - p. 108

SECTION II - SPINAL CORD, SPINAL NERVES, AUTONOMIC NERVOUS SYSTEM

FOR EACH OF THE FOLLOWING MULTIPLE CHOICE QUESTIONS SELECT THE APPROPRIATE ANSWER OR ANSWERS. ONE OR MORE, OR ALL, OF THE CHOICES MAY BE USED IN CORRECTLY ANSWERING THE QUESTION.

1116. THE AUTONOMIC CRANIAL OUTFLOW IS REPRESENTED BY PRE-GANGLIONIC VISCERAL MOTOR FIBERS WITHIN THE:
A. Oculomotor nerve
B. Facial nerve
C. Glossopharyngeal nerve
D. Vagus nerve
E. Trigeminal nerve — Ref. 8 - p. 217

1117. PARASYMPATHETIC PREGANGLIONIC CELL BODIES FOUND IN THE SUPERIOR SALIVATORY NUCLEUS OF THE PONS, SUPPLIES THE:
A. Iris of the eye
B. Lacrimal gland
C. Submandibular gland
D. Parotid gland
E. Heart — Ref. 8 - p. 221

1118. THE GREATER SPLANCHNIC NERVE ARISES BY ROOTS FROM THORACIC GANGLIA:
A. 3
B. 4
C. 5
D. 6
E. 7 — Ref. 8 - p. 221

1119. THE INTERMEDIATE PORTION OF THE FACIAL NERVE SENDS FIBERS TO THE:
A. Ciliary ganglion
B. Otic ganglion
C. Pterygopalatine ganglion
D. Submandibular ganglion
E. All of the above — Ref. 8 - p. 222

1120. SYMPATHETIC FUNCTIONS INCLUDE:
A. Miosis (constriction of the pupil)
B. Secretion of the lacrimal gland is stimulated
C. Reduced secretion of salivary gland
D. Dilates lumen of bronchi
E. Decreases heart rate — Ref. 8 - p. 227

1121. PARASYMPATHETIC FUNCTIONS INCLUDE:
A. Stimulation of GI motility
B. Relaxation of GI sphincters
C. Promotes emptying of urinary bladder
D. Stimulates secretion of adrenal medulla
E. Constricts blood vessels of extremities
Ref. 8 - p. 227

SECTION II - SPINAL CORD, SPINAL NERVES, AUTONOMIC NERVOUS SYSTEM

1122. SPINAL REFLEXES REQUIRE:
A. Peripheral receptors
B. Sensory neurons
C. Internuncial neurons
D. Motor neurons
E. Terminal effectors Ref. 8 - p. 258

1123. THE VENTRAL ROOTS OF THE UPPER LUMBAR SPINAL NERVES CONTAIN:
A. Somatic motor fibers
B. Sensory fibers
C. Postganglionic visceral afferent fibers
D. Postganglionic visceral efferent fibers
E. Myelinated preganglionic visceral efferent fibers
Ref. 7 - p. 134

1124. PROPRIOCEPTIVE FIBERS CONVEY IMPULSES FROM THE:
A. Surface of the body
B. Muscles
C. Joints
D. Semicircular canals of the ear
E. Eye Ref. 7 - p. 140

1125. THE NERVE CELLS OF THE AUTONOMIC GANGLIA ARE ALMOST ALL:
A. Unipolar
B. Pseudounipolar
C. Bipolar
D. Multipolar
E. None of the above Ref. 7 - p. 153

1126. THE WHITE RAMI CONSIST OF:
A. Visceral afferent fibers
B. Visceral efferent fibers
C. Preganglionic visceral afferent fibers
D. Preganglionic visceral efferent fibers
E. Postganglionic visceral afferent fibers
Ref. 7 - p. 159

1127. POSTGANGLIONIC THORACOLUMBAR AUTONOMIC FIBERS INCLUDE:
A. Vasoconstrictor fibers to blood vessels
B. Secretory fibers to sweat glands
C. Dilator fibers to the iris
D. Constrictor fibers to the bronchioles
E. Inhibitory fibers to the heart Ref. 7 - p. 163

1128. THE CERVICAL SPINAL CORD IN CROSS-SECTION SHOWS:
A. Circular outline
B. Larger volume of gray matter
C. Massive posterior gray column
D. Well-developed processus reticularis
E. Sulcus intermedius posterior Ref. 7 - p. 170

SECTION II - SPINAL CORD, SPINAL NERVES, AUTONOMIC NERVOUS SYSTEM

1129. THE GRAY MATTER IS COMPOSED OF:
A. Nerve cells
B. Unmyelinated axons
C. Myelinated axons
D. Capillaries
E. All of the above Ref. 7 - p. 173

1130. A REFLEX ARC CONSISTS OF:
A. An afferent nerve fiber
B. A receptor
C. Synapse in the anterior column
D. An effector
E. An efferent nerve fiber Ref. 7 - p. 178

1131. THE LARGE FIBERS IN THE MEDIAL DIVISION OF THE DORSAL ROOT ARE CONCERNED WITH:
A. Pain
B. Temperature
C. Touch
D. Pressure
E. Proprioception Ref. 7 - pp. 184-185

1132. CONCERNING THE DORSAL SPINOCEREBELLAR TRACT:
A. Begins in upper thoracic segments
B. Transmits impulses to cerebellum from muscles of trunk and legs
C. Consists of large fibers
D. Fibers arise from cells of posterior gray column
E. All of the above Ref. 7 - p. 188

1133. THE VENTRAL SPINOTHALAMIC TRACT:
A. Is a descending bundle of fibers
B. Mediates tactile sensibility
C. Crosses the median plane in the anterior white commissure
D. All of the above
E. None of the above Ref. 7 - p. 190

1134. ASCENDING TRACTS AND FIBERS OF THE SPINAL CORD INCLUDE:
A. Ventral spinothalamic tract
B. Spinotectal tract
C. Spino-olivary tract
D. Spinoreticular fibers
E. Spino-pontine fibers Ref. 7 - p. 190

1135. PATHWAYS CROSSING IN THE SPINAL CORD:
A. Pain
B. Temperature
C. Touch
D. Sensory impulses from tendons
E. Sensory impulses from joints Ref. 7 - p. 193

SECTION II - SPINAL CORD, SPINAL NERVES, AUTONOMIC NERVOUS SYSTEM

1136. THE FIBERS OF THE VENTRAL CORTICOSPINAL TRACT:
A. Cross in the medulla
B. Lie near the anterior median fissure
C. Pass into the ventral funiculus
D. All of the above
E. None of the above Ref. 7 - pp. 198-199

1137. CONCERNED WITH OPTIC AND AUDITORY REFLEXES:
A. Tectospinal tract
B. Rubrospinal tract
C. Ventral corticospinal tract
D. Olivospinal tract
E. Lateral corticospinal tract Ref. 7 - pp. 198-199

1138. RESULTING FROM HEMISECTION OF THE SPINAL CORD:
A. Loss of tactile sensibility on same side
B. Loss of sensation from muscles on opposite side
C. Loss of temperature sensation on same side
D. Loss of pain sensation on same side
E. None of the above Ref. 7 - p. 201

1139. THE TYPICAL CELLS OF THE MAMMALIAN SPINAL GANGLION ARE:
A. Unipolar
B. Bipolar
C. Tripolar
D. Multipolar
E. None of the above Ref. 7 - p. 135

1140. SKELETAL MUSCLES ARE INNERVATED BY:
A. Small, myelinated fibers
B. Small, unmyelinated fibers
C. Large, unmyelinated fibers
D. Large, myelinated fibers
E. Medium, unmyelinated fibers Ref. 7 - p. 139

1141. THE GRAY RAMI ARE COMPOSED OF:
A. Postganglionic fibers
B. Preganglionic fibers
C. Myelinated fibers
D. Unmyelinated fibers
E. Fibers to skeletal muscles Ref. 7 - p. 160

1142. WHITE SUBSTANCE OF THE SPINAL CORD CONSISTS OF:
A. Bundles of nerve fibers
B. Neuroglia fibers
C. Fibrous astrocytes
D. Neuron cell bodies
E. Ependyma Ref. 7 - p. 172

SECTION II - SPINAL CORD, SPINAL NERVES, AUTONOMIC NERVOUS SYSTEM

1143. THE CHIEF SENSATION ELICITED BY TOUCHING THE CORNEA IS:
A. Pressure
B. Heat
C. Cold
D. Pain
E. Proprioception
Ref. 7 - p. 128

1144. CUTANEOUS NERVES ASSOCIATED WITH THE UPPER EXTREMITY:
A. C-3
B. C-4
C. C-5
D. C-6
E. C-7
Ref. 7 - pp. 130-131

1145. THE DORSAL ROOT CONTAIN:
A. Only myelinated fibers
B. Only unmyelinated fibers
C. Only small myelinated fibers and large unmyelinated fibers
D. Only small unmyelinated fibers and large myelinated fibers
E. None of the above are correct statements
Ref. 7 - p. 134

1146. SATELLITE CELLS FUNCTION AS:
A. Impulse modifiers
B. Nutrient conveyors
C. Forerunners of neuroglia
D. Source of myelin
E. Function obscure
Ref. 7 - p. 136

1147. VISCERAL EFFERENT FIBERS INNERVATE:
A. Skeletal muscles
B. Glands
C. Meissner's plexus
D. Involuntary muscles
E. None of the above
Ref. 7 - p. 138

1148. THE ANTERIOR SPINOTHALAMIC TRACT CONVEYS IMPULSES OF:
A. Light touch
B. Pain
C. Heat
D. Cold
E. Stretch
Ref. 8 - p. 265

1149. THE LATERAL SPINOTHALAMIC TRACT CONVEYS IMPULSES OF:
A. Pain
B. Thermal sense
C. Proprioception
D. All of the above
E. None of the above
Ref. 8 - p. 265

SECTION II - SPINAL CORD, SPINAL NERVES, AUTONOMIC NERVOUS SYSTEM

1150. DESTRUCTION OF THE PYRAMIDAL TRACT PRODUCES:
A. Loss of voluntary movement in distal parts of the extremities
B. Loss of cremasteric reflex
C. Loss of abdominal reflex
D. Loss of thermal sensation
E. Loss of tactile sensation Ref. 8 - p. 278

1151. STIMULATION OF THE RETICULAR FORMATION OF THE BRAINSTEM CAN:
A. Influence muscle tone
B. Affect respiratory inspiration
C. Affect respiratory expiration
D. Cause pressor effects on the circulatory system
E. Inhibit voluntary movement Ref. 8 - p. 280

1152. THE VESTIBULOSPINAL TRACT IS COMPOSED OF FIBERS FROM THE:
A. Superior vestibular nucleus
B. Medial vestibular nucleus
C. Inferior vestibular nucleus
D. All of the above
E. None of the above Ref. 8 - p. 282

1153. THE RED NUCLEUS RECEIVES FIBERS FROM THE:
A. Cerebellum
B. Cerebral cortex
C. Basal ganglia
D. Pons
E. Medulla Ref. 8 - p. 286

1154. HEMISECTION OF THE SPINAL CORD (BROWN-SÉQUARD SYNDROME) CAUSES:
A. Loss of sensory impulses in the posterior white columns below the lesion, on the same side
B. Bilateral loss of pain within the area of the lesion
C. Bilateral loss of thermal sense within the area of the lesion
D. All of the above
E. None of the above Ref. 8 - p. 294

1155. IN ALL TRANSECTIONS OF THE CORD:
A. Bladder function is disturbed
B. Bowel function is unaffected
C. Extensor spasms are common
D. There is loss of superficial reflexes
E. All of the above Ref. 8 - pp. 295-296

SECTION II - SPINAL CORD, SPINAL NERVES, AUTONOMIC NERVOUS SYSTEM

ANSWER THE FOLLOWING QUESTIONS ACCORDING TO THE KEY OUTLINED BELOW:
A. If only A is correct
B. If only B is correct
C. If A and B are correct
D. If A and B are incorrect

1156. SPINAL NERVES ARE ATTACHED TO THE SPINAL CORD BY:
A. Ventral roots
B. Dorsal roots
C. Both
D. Neither Ref. 7 - p. 129

1157. CUTANEOUS BRANCHES OF SPINAL NERVES CONTAIN:
A. Fine myelinated fibers
B. Unmyelinated fibers
C. Both
D. Neither Ref. 7 - p. 133

1158. CONCERNING PREGANGLIONIC NEURONS:
A. Cell bodies lie within central nervous system
B. Axons are unmyelinated
C. Both
D. Neither Ref. 7 - p. 138

1159. EXTEROCEPTIVE FIBERS CARRY IMPULSES FROM THE:
A. Semicircular canals
B. Eye
C. Both
D. Neither Ref. 7 - p. 140

1160. PAIN IS CONVEYED BY:
A. Fine, myelinated fibers
B. Unmyelinated fibers
C. Both
D. Neither Ref. 7 - p. 141

1161. THE END ORGANS OF RUFFINI ARE ASSOCIATED WITH:
A. Cold discrimination
B. Perception of movement
C. Both
D. Neither Ref. 7 - p. 143

1162. SPECIALIZED TACTILE END ORGANS IN THE SKIN OF THE PALM:
A. End bulbs of Krause
B. Meissner's corpuscles
C. Both
D. Neither Ref. 7 - p. 146

SECTION II - SPINAL CORD, SPINAL NERVES, AUTONOMIC NERVOUS SYSTEM

1163. VISCERAL AFFERENT FIBERS ARE ASSOCIATED WITH THE:
A. 2nd sacral spinal nerve
B. 4th sacral spinal nerve
C. Both
D. Neither Ref. 7 - pp. 148-149

1164. POSTGANGLIONIC NEURONS:
A. Have their cell bodies in autonomic ganglia
B. Have mainly myelinated axons
C. Both
D. Neither Ref. 7 - p. 153

1165. AXONS OF AUTONOMIC GANGLION CELLS:
A. Usually give off numerous collaterals
B. Usually are thickly myelinated
C. Both
D. Neither Ref. 7 - p. 155

1166. EACH OF THE SPINAL NERVES HAS A:
A. White ramus from the sympathetic trunk
B. Gray ramus from the sympathetic trunk
C. Both
D. Neither Ref. 7 - p. 159

1167. PREGANGLIONIC FIBERS REACH THE CELIAC PLEXUS FROM THE:
A. White rami by way of sympathetic trunk
B. Vagus nerve
C. Both
D. Neither Ref. 7 - p. 161

1168. THE CENTRAL CANAL DIVIDES THE GRAY COMMISSURE INTO:
A. The posterior commissure
B. The anterior gray commissure
C. Both
D. Neither Ref. 7 - p. 168

1169. THE LUMBAR SPINAL CORD IN TRANSVERSE SECTION EXHIBITS:
A. An oval diameter
B. A deep sulcus intermedius posterior
C. Both
D. Neither Ref. 7 - p. 170

1170. THE POSTERIOR MEDIAN SEPTUM OF THE SPINAL CORD IS COMPOSED OF:
A. Neuroglia
B. Pia mater
C. Both
D. Neither Ref. 7 - p. 172

SECTION II - SPINAL CORD, SPINAL NERVES, AUTONOMIC NERVOUS SYSTEM

1171. THE NUCLEUS DORSALIS:
A. Extends from last cervical to 1st or 2nd lumbar segments
B. Found in the posterior column, a group of large cells
C. Both
D. Neither Ref. 7 - p. 176

1172. OBSERVED FOLLOWING SPINAL TRANSECTION:
A. Immediate response is "shock"
B. As recovery occurs, reflexes are diminished
C. Both
D. Neither Ref. 7 - p. 181

1173. THE DORSOLATERAL FASCICULUS (TRACT OF LISSAUER):
A. Is composed of unmyelinated and fine myelinated fibers
B. Lies between the apex of the posterior column and periphery of cord
C. Both
D. Neither Ref. 7 - p. 187

1174. SECTION OF THE LATERAL SPINOTHALAMIC TRACT, WHEN MADE ON ONE SIDE:
A. Results in analgesia on the same side of the body up to the caudal level of the first segment below the lesion
B. The analgesia abolishes visceral pain
C. Both
D. Neither Ref. 7 - p. 191

1175. THE VENTRAL CORTICOSPINAL TRACT IS FORMED BY:
A. Corticospinal fibers which cross in the medulla
B. Fibers which pass directly to skeletal muscles
C. Both
D. Neither Ref. 7 - pp. 198-199

FOLLOWING EACH GROUP OF FIVE CHOICES THERE ARE EITHER FIVE STATEMENTS OR FIVE ITEMS. MATCH THE STATEMENTS OR ITEMS WITH THE MOST APPROPRIATE CHOICE GIVEN. EACH CHOICE MAY BE USED ONLY ONCE:

A. Free nerve endings
B. Corpuscles of Meissner
C. End bulbs of Krause
D. Pacinian corpuscles
E. Corpuscles of Ruffini

1176. ___ Tactile end organs located in corium or subepidermal connective tissue

1177. ___ End among epithelial cells, pain receptors

1178. ___ Proprioceptive function

1179. ___ Heat perception

1180. ___ Cold perception Ref. 7 - pp. 142-143

SECTION II - SPINAL CORD, SPINAL NERVES, AUTONOMIC NERVOUS SYSTEM

A. Dorsal spinocerebellar tract
B. Ventral spinocerebellar tract
C. Spinotectal tract
D. Spino-olivary tract
E. Spinoreticular fibers

1181. ___ Consists of fibers arising in posterior gray column

1182. ___ Direct cerebellar tract of Flechsig

1183. ___ Situated at periphery of lateral funiculus

1184. ___ From animal experiments it appears that tract carries impulses to medial (vermal) portion of anterior lobe of cerebellum

1185. ___ May be route of exteroceptive impulses to cerebellum

Ref. 7 - pp. 188-189, 193

SECTION II - SPINAL CORD, SPINAL NERVES, AUTONOMIC NERVOUS SYSTEM

IDENTIFY THE TRACTS IN THE OUTLINE OF THE SPINAL CORD; FIVE ASCENDING TRACT ON THE RIGHT AND FIVE DESCENDING TRACTS ON THE LEFT.

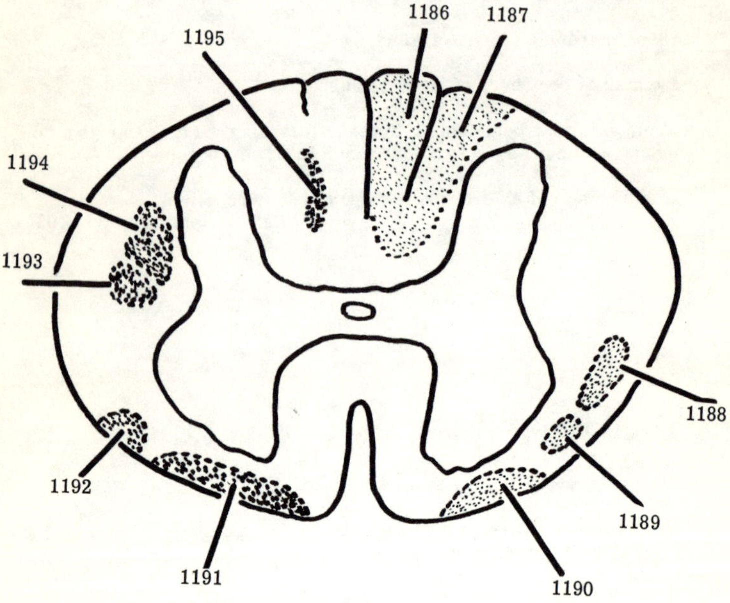

A. Ventral spinothalamic tract
B. Fasciculus interfascicularis
C. Lateral spinothalamic tract
D. Fasciculus proprius
E. Fasciculus gracilis
F. Vestibulospinal tract
G. Spinotectal tract
H. Rubrospinal tract
I. Dorsolateral fasciculus
J. Dorsal spinocerebellar tract
K. Fasciculus cuneatus
L. Fasciculus septomarginalis
M. Olivospinal tract
N. Ventral corticospinal tract
O. Lateral corticospinal tract

Ref. 7 - p. 198

SECTION III - MEDULLA OBLONGATA, PONS, MESENCEPHALON, CRANIAL NERVES

FOR EACH OF THE FOLLOWING MULTIPLE CHOICE QUESTIONS SELECT THE ONE APPROPRIATE ANSWER.

1196. ALL OF THE FOLLOWING ARE CHARACTERISTIC OF THE MEDULLA, EXCEPT FOR:
A. Development of third ventricle
B. Decussation of the pyramids
C. Development of cranial nerve nuclei
D. Appearance of complex groups of relay nuclei, part of which belong to the reticular formation
E. Gradual replacement of spinal fibers in zone of Lissauer by descending spinal trigeminal fibers Ref. 8 - p. 298

1197. A SECTION THROUGH THE ADULT MEDULLA, IN THE MID-OLIVARY REGION, WOULD NOT DEMONSTRATE THE:
A. Hypoglossal nucleus
B. Pyramid
C. Lateral reticular nucleus
D. Nucleus ambiguus
E. Red nucleus Ref. 8 - p. 315

1198. ALL OF THE FOLLOWING MEDULLARY NUCLEI CONTRIBUTE TO THE FORMATION OF THE INFERIOR CEREBELLAR PEDUNCLE, EXCEPT FOR THE:
A. Lateral reticular nucleus
B. Accessory cuneate nucleus
C. Vagus nucleus
D. Arcuate nucleus
E. Nucleus of Roller Ref. 8 - p. 319

1199. AXONS FROM THE NUCLEUS AMBIGUUS INNERVATE MUSCLES OF THE:
A. Eye
B. Tongue
C. Larynx
D. Ear
E. Palate Ref. 8 - p. 322

1200. ONE OF THE CRANIAL NERVES CONTRIBUTING TO THE FORMATION OF FASCICULUS SOLITARIUS IS THE:
A. Spinal accessory
B. Vagus
C. Hypoglossal
D. Trigeminal
E. Optic Ref. 8 - p. 323

1201. CELLS OF THE RESPIRATORY CENTER ARE FOUND IN THE:
A. Pyramid
B. Medial lemniscus
C. Inferior olive
D. Reticular formation
E. Corticospinal tract Ref. 8 - p. 324

SECTION III - MEDULLA OBLONGATA, PONS, MESENCEPHALON, CRANIAL NERVES

1202. AT THE LEVEL OF THE UPPER MEDULLA, AND AREA OF JUNCTION OF MEDULLA AND PONS, THE CRANIAL NERVE NUCLEUS FOUND IS THE:
A. Hypoglossal
B. Dorsal motor nucleus of the vagus
C. Glossopharyngeal
D. Trigeminal
E. Spinal accessory Ref. 8 - p. 330

1203. WHICH OF THE FOLLOWING IS HOMOLOGOUS WITH THE SPINAL GANGLIA:
A. Semilunar ganglion
B. Geniculate ganglion
C. Vestibular ganglion
D. All of the above
E. None of the above Ref. 7 - p. 202

1204. CONCERNING THE MEDIAL LEMNISCUS OF THE MEDULLA OBLONGATA:
A. Lies close to the median raphe
B. Lateral to the inferior olivary nucleus
C. Ventral to the pyramids
D. All of the above
E. None of the above Ref. 7 - p. 209

1205. THE INFERIOR CEREBELLAR PEDUNCLE IS COMPOSED OF THE:
A. Olivocerebellar fibers
B. Dorsal spinocerebellar tract
C. Dorsal external arcuate fibers
D. Fibers from the lateral reticular nucleus
E. All of the above Ref. 7 - p. 213

1206. THE NUCLEUS AMBIGUUS GIVES ORIGIN TO MOTOR FIBERS THAT RUN THROUGH THE:
A. Vagus, trigeminal and facial nerves
B. Glossopharyngeal, vagus and accessory nerves
C. Vagus, hypoglossal and facial nerves
D. Facial, abducens and oculomotor nerves
E. Trigeminal, abducens and facial nerves
Ref. 7 - p. 216

1207. THE NUCLEUS OF THE TRACTUS SOLITARIUS CONTAINS CELLS ON WHICH AFFERENT FIBERS OF VARIOUS CRANIAL NERVES TERMINATE:
A. Abducens, facial, vagus
B. Facial, vagus, accessory
C. Facial, glossopharyngeal, vagus
D. Glossopharyngeal, vagus, accessory
E. Trigeminal, facial, vagus Ref. 7 - p. 216

SECTION III - MEDULLA OBLONGATA, PONS, MESENCEPHALON, CRANIAL NERVES

1208. THE NUCLEI PONTIS ARE CONTINUOUS WITH THE ___________ OF THE MEDULLA OBLONGATA:
A. Arcuate nucleus
B. Lateral reticular nucleus
C. Hypoglossal nucleus
D. Vagal nucleus
E. Spinal vestibular nucleus Ref. 7 - p. 219

1209. FIBERS FROM THE DORSAL COCHLEAR NUCLEUS JOIN FIBERS FROM THE VENTRAL COCHLEAR NUCLEUS IN THE:
A. Nucleus of the vestibular nerve
B. Trapezoid body
C. Medial longitudinal fasciculus
D. Medial lemniscus
E. None of the above Ref. 7 - p. 220

1210. THE VESTIBULOCEREBELLAR FASCICULUS TERMINATES IN THE ___________ OF THE CEREBELLUM:
A. Nodulus
B. Uvula
C. Lingula
D. Fastigial nuclei
E. All of the above Ref. 7 - p. 221

1211. THE FACIAL COLLICULUS IS FORMED BY THE:
A. Abducens and facial
B. Facial and vagus
C. Facial and glossopharyngeal
D. Facial and trigeminal
E. Facial and hypoglossal Ref. 7 - p. 222

1212. NUCLEAR MASSES ASSOCIATED WITH THE TRIGEMINAL NERVE:
A. Nucleus of spinal tract
B. Main sensory nucleus
C. Motor nucleus
D. Mesencephalic nucleus
E. All of the above Ref. 7 - p. 225

1213. THE RED NUCLEUS IS LOCATED IN THE PATH OF THE:
A. Medial lemniscus
B. Medial longitudinal fasciculus
C. Fasciculus gracilis
D. Superior cerebellar peduncle
E. Inferior cerebellar peduncle Ref. 7 - p. 230

1214. THE TRACTUS SOLITARIUS IS FORMED BY SENSORY FIBERS OF CRANIAL NERVES:
A. V, VII, X
B. VI, VIII, IX
C. VII, IX, X
D. X, XI, XII
E. V, VI, VII Ref. 7 - p. 203

SECTION III - MEDULLA OBLONGATA, PONS, MESENCEPHALON, CRANIAL NERVES

1215. THE DORSAL EXTERNAL ARCUATE FIBERS TERMINATE IN THE:
A. Cerebellum
B. Pons
C. Medulla oblongata
D. Spinal cord
E. Hypothalamus Ref. 7 - p. 210

1216. THE NUCLEUS AMBIGUUS IS ASSOCIATED WITH STRIATED MUSCULATURE OF THE:
A. Eye
B. Pharynx
C. Soft palate
D. Tongue
E. Mandible Ref. 7 - p. 216

1217. THE MAJORITY OF FIBERS OF THE FIBRAE PONTIS FORM A MASSIVE STRAND, KNOWN AS THE:
A. Superior cerebellar peduncle
B. Middle cerebellar peduncle
C. Inferior cerebellar peduncle
D. Olivary tract
E. Reticulospinal bundle Ref. 7 - pp. 218-219

1218. THE SUPERIOR OLIVARY NUCLEUS IS ANATOMICALLY SITUATED NEAR THE:
A. Restiform body
B. Red nucleus
C. Anterior medullary velum
D. Vermis
E. Trapezoid body Ref. 7 - p. 221

1219. THE MEDIAL LEMNISCUS IS IN PART DERIVED FROM FIBERS OF THE:
A. Gracile and cuneate nuclei of the same side
B. Gracile and cuneate nuclei of the opposite side
C. Gracile nucleus of the same side and the cuneate nucleus of opposite side
D. Gracile nucleus of the opposite side and the cuneate nucleus of the same side
E. None of the above Ref. 7 - p. 232

1220. THE NUCLEUS OF THE TROCHLEAR NERVE CONTAINS CELLS OF ORIGIN OF MOTOR FIBERS FOR THE ____________ MUSCLES OF THE EYE:
A. Rectus medialis
B. Rectus lateralis
C. Superior oblique
D. Inferior oblique
E. Palpebral Ref. 7 - p. 232

SECTION III - MEDULLA OBLONGATA, PONS, MESENCEPHALON, CRANIAL NERVES

1221. THE SUBSTANTIA NIGRA IS PART OF THE:
A. Cerebellum
B. Medulla oblongata
C. Pons
D. Mesencephalon
E. All of the above Ref. 7 - p. 233

1222. THE INFERIOR QUADRIGEMINAL BRACHIUM CONSISTS OF FIBERS FROM THE:
A. Stratum griseum
B. Superior quadrigeminal brachium
C. Inferior colliculus
D. Medial lemniscus
E. Lateral lemniscus Ref. 7 - p. 234

1223. FUNCTIONAL COMPONENTS IN THE CRANIAL NERVES INCLUDE:
A. Somatic efferent, special somatic afferent, special visceral efferent
B. General somatic afferent, general visceral efferent, general visceral afferent
C. General visceral efferent, special visceral afferent, somatic efferent
D. All of the above
E. None of the above Ref. 7 - p. 240

1224. THE VISCERAL AFFERENT COLUMN INVOLVES THE FOLLOWING CRANIAL NERVES:
A. Facial, trigeminal, abducens
B. Facial, trigeminal, vagus
C. Glossopharyngeal, facial, vagus
D. Glossopharyngeal, facial, trigeminal
E. None of the above Ref. 7 - p. 241

1225. THE MOTOR NUCLEUS OF THE FACIAL NERVE IS LOCATED IN THE:
A. Spinal cord
B. Medulla oblongata
C. Pons
D. Mesencephalon
E. Thalamus Ref. 7 - p. 246

1226. THE EDINGER-WESTPHAL NUCLEUS IS LOCATED IN THE ROSTRAL PART OF THE NUCLEUS OF THE ____________ NERVE:
A. Ophthalmic
B. Oculomotor
C. Optic
D. Abducens
E. Trochlear Ref. 7 - p. 249

SECTION III - MEDULLA OBLONGATA PONS, MESENCEPHALON, CRANIAL NERVES

1227. ALL OF THE GENERAL AND SPECIAL AFFERENT FIBERS, WHETHER CONTAINED IN THE SEVENTH, NINTH OR TENTH NERVES, ENTER THE:
A. Medial lemniscus
B. Lateral lemniscus
C. Medial longitudinal fasciculus
D. Tractus solitarius
E. Fasciculus gracilis
Ref. 7 - p. 252

1228. INJURY OF THE PYRAMIDAL TRACT ANYWHERE ABOVE THE DECUSSATION MAY CAUSE:
A. Paralysis of the contralateral extremities
B. Paralysis of the ipsilateral extremities
C. Loss of pain and touch on the contralateral side
D. Loss of thermal sensation on the contralateral side
E. Loss of proprioception on the contralateral side
Ref. 8 - p. 301

1229. IN THE POSTERIOR WHITE COLUMNS OF THE MEDULLA,NUCLEAR MASSES APPEAR IN THE FASCICULI:
A. Solitarius and arcuatus
B. Cuneatus and gracilis
C. Gracilis and solitarius
D. Solitarius and cuneatus
E. None of the above
Ref. 8 - p. 302

1230. MYELINATED FIBERS FROM NUCLEI GRACILIS AND CUNEATUS RUN VENTROMEDIALLY AROUND THE CENTRAL GRAY; THEY ARE KNOWN AS:
A. Fasciculi cuneatus and gracilis
B. Pyramidal decussation
C. Internal arcuate fibers
D. Lateral corticospinal tract
E. Posterior spinocerebellar tract
Ref. 8 - p. 307

1231. INJURY TO THE MEDIAL LEMNISCUS CAUSES:
A. Paralysis of extremities on the opposite side
B. Tactile deficits on the opposite side
C. Pain deficits on the same side
D. Temperature deficits on the same side
E. Proprioceptive deficits on the same side
Ref. 8 - p. 309

1232. GENERAL SOMATIC AFFERENT FIBERS DESCEND IN THE SPINAL TRIGEMINAL TRACT FROM THE:
A. Facial and trigeminal nerves
B. Trigeminal, glossopharyngeal and hypoglossal nerves
C. Hypoglossal and vagus nerves
D. Hypoglossal and facial nerves
E. Vagus, glossopharyngeal and facial nerves
Ref. 8 - p. 311-312

SECTION III - MEDULLA OBLONGATA, PONS, MESENCEPHALON, CRANIAL NERVES

1233. ROSTRAL TO THE OBEX ON EACH SIDE OF THE FOURTH VENTRICLE IS THE:
A. Lateral reticular nucleus
B. Hypoglossal nucleus
C. Area postrema
D. Arcuate nucleus
E. Pyramid Ref. 8 - p. 314

1234. ON THE ANTERIOR ASPECT OF THE PYRAMID IS THE:
A. Arcuate nucleus
B. Pontobulbar nucleus
C. Lateral reticular nucleus
D. Hypoglossal nucleus
E. Paramedian reticular nuclei Ref. 8 - p. 314

1235. THE ROSTRAL ENLARGED PORTION OF THE NUCLEUS SOLITARIUS IS REFERRED TO AS THE:
A. Respiratory center
B. Cardiac center
C. Vomiting nucleus
D. Gustatory nucleus
E. Sinus nerve Ref. 8 - p. 324

SELECT THE ITEM FROM COLUMN I WHICH IS UNRELATED TO THE FOUR OTHER ITEMS IN THIS COLUMN, CIRCLE THIS LETTER. IDENTIFY THE ITEM IN COLUMN II TO WHICH THE FOUR ITEMS IN COLUMN I ARE RELATED, CIRCLE THIS NUMBER.

	COLUMN I	COLUMN II
1236.	A. Basis pedunculi	1. Mesencephalon
	B. Nucleus gracilis	2. Pons
	C. Decussation of pyramids	3. Medulla
	D. Nucleus of hypoglossal nerve	
	E. Inferior olivary nucleus	Ref. 7 - p. 205
1237.	A. Nucleus gracilis	1. White matter
	B. Nucleus cuneatus	2. Gray matter
	C. Reticular substance	3. Mixture of white and gray matter
	D. Inferior olivary nucleus	
	E. Nucleus of hypoglossal nerve	Ref. 7 - p. 206
1238.	A. Olivocerebellar fibers	1. Inferior cerebellar peduncle
	B. Dorsal spinocerebellar tract	2. Medial lemniscus
	C. Dorsal external arcuate fibers	3. Pyramids
	D. Longitudinal fasciculi	
	E. Fibers from lateral reticular nucleus	Ref. 7 - p. 213

SECTION III - MEDULLA OBLONGATA, PONS, MESENCEPHALON, CRANIAL NERVES

COLUMN I	COLUMN II
1239. A. Contains cells of origin of motor fibers	1. Dorsal motor nucleus of vagus
B. Fibers emerge at lateral border of pyramid	2. Nucleus of hypoglossal nerve
C. On lateral side is nucleus intercalatus	3. Nucleus of tractus solitarius
D. Fibers pass through reticular formation	
E. Occupies ala cinera of rhomboid fossa	Ref. 7 - p. 216
1240. A. Continuous with arcuate nucleus of medulla oblongata	1. Lateral vestibular nucleus
B. Surrounds fourth ventricle	2. Trapezoid body
C. Axons continuous with fibrae pontis	3. Nuclei pontis
D. Fibers of corticopontine tract terminate here	
E. Collaterals from medial lemniscus also arborize here	Ref. 7 - p. 219
1241. A. Covered by pars basilis pontis	1. Trapezoid body
B. Largest gray mass of the medulla	2. Vestibulocerebellar fasciculus
C. Fibers associated with terminal nuclei of cochlear nerve	3. Medial longitudinal fasciculus
D. In caudal pons seen as a bundle in the reticular formation	
E. Fibers associated with superior olivary nucleus	Ref. 7 - p. 220
1242. A. Extends from near floor of 3rd ventricle to spinal cord	1. Inferior cerebellar peduncle
B. Concerned with reflex control of movements of head and eyes	2. Medial longitudinal fasciculus
C. Major central auditory pathway	3. Medial lemniscus
D. Numerous fibers derived from vestibular nuclei	
E. Also has fibers from interstitial nucleus	Ref. 7 - pp. 221-222
1243. A. Nucleus of inferior colliculus	1. Medulla oblongata
B. Lateral lemniscus	2. Pons
C. Interpeduncular fossa	3. Mesencephalon
D. Arcuate nucleus	
E. Substantia nigra	Ref. 7 - p. 228

SECTION III - MEDULLA OBLONGATA, PONS, MESENCEPHALON, CRANIAL NERVES

COLUMN I	COLUMN II
1244. A. Glossopharyngeal nerve	1. General somatic afferent fibers
B. Oculomotor nerve	2. Special visceral afferent fibers
C. Trigeminal nerve	3. General visceral efferent fibers
D. Vagus nerve	
E. Trochlear nerve	Ref. 7 - pp. 271-272

1245. A. General somatic afferent	1. Facial nerve and nervus intermedius
B. General visceral afferent fibers	2. Trigeminal nerve
C. Special visceral afferent fibers	3. Oculomotor nerve
D. Special visceral efferent fibers	
E. General visceral efferent fibers	Ref. 7 - pp. 270-273

FOLLOWING EACH GROUP OF FIVE CHOICES THERE ARE EITHER FIVE STATEMENTS OR FIVE ITEMS. MATCH THE STATEMENTS OR ITEMS WITH THE MOST APPROPRIATE CHOICE GIVEN. EACH CHOICE MAY BE USED ONLY ONCE:

A. Optic nerve
B. Trochlear nerve
C. Trigeminal nerve
D. Vagus nerve
E. Hypoglossal nerve

1246. ___ Contains proprioceptive fibers from cells of origin in mesencephalic nucleus of V

1247. ___ Superficial origin from anterior medullary velum

1248. ___ Has special visceral efferent fibers from nucleus ambiguus

1249. ___ Superficial origin from anterior lateral sulcus of medulla oblongata

1250. ___ Not a true nerve Ref. 7 - pp. 270-273

A. Superior olivary nucleus
B. Superior cerebellar peduncle
C. Medial longitudinal fasciculus
D. Superior quadrigeminal bodies
E. Inferior cerebellar peduncle

1251. ___ Runs from dentate nucleus of cerebellum to red nucleus of mesencephalon

1252. ___ Composed of olivocerebellar fibers and dorsal spinocerebellar tract

1253. ___ Majority of afferent fibers from optic tract

1254. ___ Close relation to the nuclei for the motor nerves of the eye

1255. ___ Located in ventrolateral portion of reticular formation of pons in close relation to trapezoid body Ref. 7 - pp. 213, 221, 225, 231

SECTION III - MEDULLA OBLONGATA, PONS, MESENCEPHALON, CRANIAL NERVES

IDENTIFY THE STRUCTURES OF THE MEDULLA OBLONGATA IN A SECTION MADE AT THE LEVEL OF THE OLIVE:

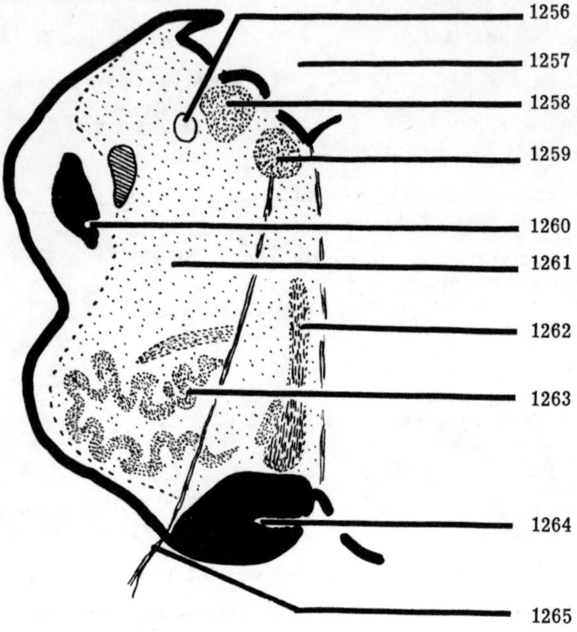

A. Dorsal motor nucleus of the vagus
B. Medial lemniscus
C. Medial accessory olivary nucleus
D. Fibers of hypoglossal nerve
E. Tractus solitarius
F. Nucleus of hypoglossal nerve
G. Reticular substance
H. Lateral lemniscus
I. Inferior olivary nucleus
J. Fourth ventricle
K. Pyramid, corticospinal tract
L. Third ventricle
M. Fibers of abducens nerve
N. Nucleus of spinal tract of trigeminal nerve
O. Spinal tract of trigeminal nerve

Ref. 7 - p. 205

SECTION IV - CEREBELLUM, DIENCEPHALON, RHINENCEPHALON, CEREBRUM

FOR EACH OF THE FOLLOWING MULTIPLE CHOICE QUESTIONS SELECT THE APPROPRIATE ANSWER OR ANSWERS. ONE OR MORE, OR ALL, OF THE CHOICES MAY BE USED IN CORRECTLY ANSWERING THE QUESTION.

1266. BY THREE CEREBELLAR PEDUNCLES, THE CEREBELLUM IS ATTACHED TO THE:
A. Medulla
B. Spinal cord
C. Pons
D. Cerebrum
E. Midbrain — Ref. 8 - p. 406

1267. CELL TYPES FOUND IN THE MOLECULAR LAYER OF THE CEREBELLUM:
A. Basket cells
B. Purkinje cells
C. Granule cells
D. Mossy cell
E. Outer stellate cells — Ref. 8 - p. 407

1268. "CEREBELLAR ISLANDS" OR GLOMERULI OF THE CEREBELLUM ARE FORMED BY:
A. One mossy fiber rosette
B. Cell bodies of Purkinje cells
C. Terminals of Golgi cell axons
D. Cell bodies of outer stellate cells
E. None of the above — Ref. 8 - p. 412

1269. IN THE WHITE MATTER OF THE CEREBELLUM ARE THE FOLLOWING NUCLEI:
A. Nucleus caudatus
B. Nucleus fastigii
C. Nucleus dentatus
D. Nucleus emboliformis
E. Nucleus globosus — Ref. 8 - pp. 416-418

1270. THE CEREBELLUM IS ASSOCIATED WITH:
A. Integration of proprioception and reflex activity
B. Tactile sense appreciation
C. Coordination of somatic motor activity
D. Regulation of muscle tone
E. Regulation of cardiac activity — Ref. 8 - p. 433

1271. LESIONS OF THE NEOCEREBELLUM ARE MANIFESTED BY:
A. Hypotonia (flabby muscles)
B. Asynergia (severe disorders of movement)
C. Dysmetria (inability to guage distances correctly)
D. Past-pointing
E. None of the above — Ref. 8 - p. 434

SECTION IV - CEREBELLUM, DIENCEPHALON, RHINENCEPHALON, CEREBRUM

1272. THE TREMOR ASSOCIATED WITH CEREBELLAR LESIONS:
- A. Occurs at rest and during sleep
- B. Has a coarse, irregular quality
- C. Occurs ipsilateral to the lesion
- D. Is seen predominantly in the extremities
- E. Is known as a "resting" tremor Ref. 8 - p. 434

1273. CEREBELLAR ATAXIA:
- A. Is an asynergic disturbance
- B. Is associated with neocerebellar lesions
- C. Especially involves the axial muscles, muscles of the shoulder and pelvic girdles
- D. All of the above
- E. None of the above Ref. 8 - p. 435

1274. LESIONS OF THE POSTERIOR CEREBELLAR VERMIS (ARCHI-CEREBELLAR SYNDROME) ARE CHARACTERIZED BY:
- A. Patient unsteady when standing
- B. Gait resembles the stagger of a drunkard
- C. Marked hypotonia
- D. Resting tremor
- E. Highest incidence in the elderly Ref. 8 - p. 435

1275. THE EPITHALAMUS INCLUDES THE:
- A. Pituitary gland
- B. Pineal body
- C. Habenular trigones
- D. Roof of the fourth ventricle
- E. All of the above Ref. 8 - p. 458

1276. THE ANTERIOR LIMB OF THE INTERNAL CAPSULE CONTAINS THE:
- A. Prefrontal corticopontine tract
- B. Corticospinal fibers
- C. Superior thalamic radiation
- D. Fibers from the orbital cortex to the hypothalamus
- E. Corticobulbar fibers Ref. 8 - p. 476

1277. HOMONYMOUS VISUAL DEFECTS ARE ASSOCIATED WITH LESIONS OF THE:
- A. Optic tract
- B. Lateral geniculate body
- C. Optic radiation
- D. Visual cortex
- E. None of the above Ref. 8 - p. 481

1278. THE BASAL GANGLIA INCLUDE:
- A. Caudate nucleus
- B. Putamen
- C. Globus pallidus
- D. Red nucleus
- E. Circinate nucleus Ref. 8 - p. 498

SECTION IV - CEREBELLUM, DIENCEPHALON, RHINENCEPHALON, CEREBRUM

1279. THE PUTAMEN:
A. Is the largest of the basal ganglia
B. Is the most lateral part of the basal ganglia
C. Lies between external capsule and lateral medullary lamina of the globus pallidus
D. All of the above
E. None of the above — Ref. 8 - p. 502

1280. DISTURBANCES ASSOCIATED WITH DISEASES OF THE EXTRA-PYRAMIDAL SYSTEM:
A. Dyskinesia
B. Tremor
C. Athetosis
D. Chorea
E. Ballism — Ref. 8 - p. 513

1281. NUCLEI OF THE CEREBELLUM INCLUDE:
A. Red nucleus
B. Vestibular nucleus
C. Dentate nucleus
D. Emboliform nucleus
E. Globose nucleus — Ref. 7 - p. 277

1282. THE INFERIOR CEREBELLAR PEDUNCLE CONSISTS OF THE FOLLOWING ASCENDING FIBERS:
A. Dorsal spinocerebellar tract
B. Ventral spinocerebellar tract
C. Olivocerebellar tract
D. Dorsal external arcuate fibers
E. Tectocerebellar tract — Ref. 7 - p. 278

1283. FUNCTIONS OF THE CEREBELLUM:
A. Deals with synergy of the skeletal muscles
B. Deals with conscious sensation
C. Cerebellar effect in certain visceral mechansims
D. All of the above
E. None of the above — Ref. 7 - p. 285

1284. STIMULATION OF THE CEREBELLUM RESULTS IN:
A. Contraction of skeletal muscles in specific regions
B. Disturbance of balance and posture
C. Inhibition of excessive tone in decerebrate rigidity
D. Increased perspiration and salivation
E. None of the above — Ref. 7 - p. 287

1285. SYMPTOMS RESULTING FROM INJURY TO THE NEOCEREBELLUM:
A. Nystagmus
B. Dysmetria
C. Tremor
D. Ataxia
E. Hypotonia — Ref. 7 - p. 290

SECTION IV - CEREBELLUM, DIENCEPHALON, RHINENCEPHALON, CEREBRUM

1286. THE THALAMUS INCLUDES THE:
A. Stratum zonale
B. External medullary lamina
C. Nucleus semilunaris
D. All of the above
E. None of the above Ref. 7 - p. 293

1287. NUCLEI OF THE THALAMUS ASSOCIATED WITH OLFACTORY CONNECTIONS:
A. Anterior thalamic nuclei
B. Nuclei of the midline
C. Medial nuclei
D. Lateral nuclear mass
E. Posterior thalamic nuclei Ref. 7 - pp. 294-295

1288. THE MEDIAL GENICULATE BODY IS ASSOCIATED WITH:
A. Olfaction
B. Vision
C. Hearing
D. Taste
E. Balance Ref. 7 - p. 297

1289. LESIONS IN THE LATERAL NUCLEUS OF THE THALAMUS OFTEN CAUSE:
A. Impairment of sensation on same side of body
B. Intractable pain on same side of body
C. Impairment of sensation on opposite side of body
D. Intractable pain on opposite side of body
E. A combination of (A) and (D) Ref. 7 - p. 299

1290. THE EPITHALAMUS INCLUDES THE:
A. Pineal body
B. Ansa lenticularis
C. Stria medullaris
D. Habenular trigone
E. Mammillary body Ref. 7 - p. 301

1291. THE OPTIC CHIASMA AND TUBER CINEREUM ARE PART OF THE:
A. Hypothalamus
B. Epithalamus
C. Subthalamus
D. (B) and (C)
E. (A) and (B) Ref. 7 - p. 302

1292. HYPOTHALAMIC LESIONS OFTEN CAUSE:
A. Somnolence
B. Disturbances of temperature regulation
C. Disturbances of fat metabolism
D. Dysmetria
E. Ataxia Ref. 7 - pp. 308-311

SECTION IV - CEREBELLUM, DIENCEPHALON, RHINENCEPHALON, CEREBRUM

1293. THE INTERNAL CAPSULE CONTAINS FIBERS OF THE:
A. Rubrospinal tract
B. Tectospinal tract
C. Corticopontile tract
D. Corticospinal tract
E. Corticobulbar tract Ref. 7 - p. 321

1294. COMMISSURES JOINING THE CEREBRAL HEMISPHERES INCLUDE:
A. Anterior medullary velum
B. Corpus callosum
C. Anterior commissure
D. Hippocampal commissure
E. Striae terminalis Ref. 7 - p. 322

1295. CONNECTING THE ORBITAL GYRI OF THE FRONTAL LOBE WITH THE ROSTRAL PART OF THE TEMPORAL LOBE:
A. Superior occipitofrontal fasciculus
B. Uncinate fasciculus
C. Cingulum
D. Superior longitudinal fasciculus
E. Inferior occipitofrontal fasciculus Ref. 7 - p. 325

1296. CONCERNING THE AMYGDALOID BODY:
A. Found near the inferior horn of the lateral ventricle
B. Main efferent pathway is stria terminalis
C. Function is obscure
D. All of the above
E. None of the above Ref. 7 - p. 333

1297. THE HIPPOCAMPUS IS ASSOCIATED WITH:
A. Taste
B. Smell
C. Sight
D. Balance
E. Hearing Ref. 7 - p. 334

1298. THE TWO LATERAL VENTRICLES ARE SEPARATED BY THE:
A. Fornix
B. Hippocampus
C. Amygdaloid body
D. Cingula
E. Septum pallucidum Ref. 7 - p. 337

1299. THE MEDIAL FOREBRAIN BUNDLE CONNECTS THE VENTROMEDIAL OLFACTORY CENTERS WITH THE:
A. Hypothalamus
B. Epithalamus
C. Finely myelinated fibers of the hippocampus
D. Preoptic area rostral and dorsal to the optic chiasma
E. None of the above Ref. 7 - p. 345

SECTION IV - CEREBELLUM, DIENCEPHALON, RHINENCEPHALON, CEREBRUM

1300. THE STRIA TERMINALIS JOINS THE:
A. Cerebral hemispheres posteriorly
B. Amygdaloid nuclei
C. Hippocampal nuclei
D. Olfactory bulbs
E. Otic ganglia Ref. 7 - p. 346

1301. THE CEREBELLUM RECEIVES AFFERENT IMPULSES FROM THE:
A. Spinal cord
B. Medulla oblongata
C. Pons
D. Mesencephalon
E. Cerebrum Ref. 7 - p. 277

1302. THE SUPERIOR CEREBELLAR PEDUNCLE CONSISTS OF EFFERENT FIBERS DERIVED FROM THE:
A. Red nucleus
B. Amygdaloid nucleus
C. Cuneate nucleus
D. Globose nucleus
E. None of the above Ref. 7 - p. 278

1303. THE CELLS OF PURKINJE OF THE CEREBELLUM:
A. Have small spherical cell bodies
B. Are arranged in a continuous sheet
C. Have no nuclei
D. Have no dendrites
E. All of the above Ref. 7 - p. 283

1304. CONCERNING THE STRUCTURE OF THE THALAMUS:
A. Consists chiefly of white matter
B. Dorsal surface covered by stratum zonale
C. On lateral surface is the external medullary lamina
D. All of the above
E. None of the above Ref. 7 - p. 293

1305. A RIGHT-HANDED MAN WITH AN ABLATION OF THE RIGHT TEMPORAL LOBE:
A. Will be completely deaf
B. 50% deafness results
C. 25% deafness results
D. Little or no impairment of hearing results
E. None of the above Ref. 7 - p. 371

SECTION IV - CEREBELLUM, DIENCEPHALON, RHINENCEPHALON, CEREBRUM

ANSWER THE FOLLOWING QUESTIONS USING THE KEY AS OUTLINED BELOW:

A. If 1, 2 and 3 are correct
B. If 1 and 3 are correct
C. If 2 and 4 are correct
D. If all are correct
E. If some combination other than the above are correct
F. If all are incorrect

1306. THE ARCHIPALLIUM IS REPRESENTED BY THE:
1. Hippocampal formation
2. Dentate gyrus
3. Fasciolar gyrus
4. Indusium griseum — Ref. 8 - p. 518

1307. LAYERS OF THE DENTATE GYRUS:
1. Molecular layer
2. Granular layer
3. Polymorphic layer
4. Purkinje layer — Ref. 8 - p. 533

1308. PRINCIPAL CELL TYPES FOUND IN THE CEREBRAL CORTEX:
1. Pyramidal cells
2. Stellate cells
3. Fusiform cells
4. Purkinje cells — Ref. 8 - p. 543

1309. CELL LAYERS OF THE CEREBRAL CORTEX:
1. Mitochondrial layer
2. Betz cell layer
3. Purkinje cell layer
4. Unipolar cell layer — Ref. 8 - p. 544

1310. ESTABLISHED PRIMARY SENSORY AREAS IN THE CEREBRAL CORTEX:
1. Visual
2. Tactile
3. Auditory
4. Thermal — Ref. 8 - p. 555

1311. THE PREMOTOR AREA (AREA 6):
1. Lies in front of the motor area
2. Is composed of large well-formed giant cells of Betz
3. Is associated with gustatory sensation
4. Is part of the occipital lobe — Ref. 8 - pp. 573-574

1312. THE INFERIOR CEREBELLAR PEDUNCLE CONTAINS:
1. Descending fibers of olivocerebellar tract
2. Ascending fibers of dorsal spinocerebellar tract
3. Descending fibers of ventral external arcuate fibers
4. Ascending fibers of dorsal external arcuate fibers

Ref. 7 - p. 278

SECTION IV - CEREBELLUM, DIENCEPHALON, RHINENCEPHALON, CEREBRUM

1313. FEATURES OF MOLECULAR LAYER OF CEREBELLUM:
1. Rich in nerve cells
2. Contains dendrites of Purkinje cells
3. Basket cells found here
4. Contains large numbers of stellate cells

Ref. 7 - p. 284

1314. RELATIVE TO THE STRUCTURE OF THE THALAMUS:
1. Consists chiefly of gray matter
2. Dorsal surface covered by stratum zonale
3. Lateral surface reveals external medullary lamina
4. Medial surface covered by layer of central matter

Ref. 7 - p. 293

1315. AMONG THE DEEPLY PLACED MASSES OF GRAY MATTER WITHIN THE CEREBRAL HEMISPHERES:
1. Caudate nucleus
2. Lentiform nucleus
3. Amygdaloid nucleus
4. Claustrum

Ref. 7 - p. 312

1316. PATIENTS WITH DISEASED BASAL GANGLIA EXHIBIT:
1. The Parkinsonian syndrome
2. Athetosis
3. Chorea
4. Aphasia

Ref. 7 - p. 317

1317. THE POSTERIOR LIMB OF THE INTERNAL CAPSULE:
1. Intervenes between the lentiform nucleus and septum pallucidum
2. Consists of 2 parts: lenticulothalamic and retrolenticular portions
3. All fibers run forward
4. Is a broad band of gray matter

Ref. 7 - pp. 319-320

1318. THE HIPPOCAMPUS IS:
1. A higher order of the olfactory system
2. Fused with the parahippocampal gyrus
3. Formed by highly specialized cortex
4. Part of the roof of the inferior horn of the lateral ventricle

Ref. 7 - pp. 334-335

1319. THE FASCIA DENTATA:
1. Is associated with speech
2. Belongs to the archipallium
3. Is found only in monkeys
4. Is closely related to the hippocampus

Ref. 7 - p. 344

1320. "UNCINATE FITS" ARE CHARACTERIZED BY:
1. Transient loss of vision
2. Olfactory hallucinations
3. A sense of unreality
4. Basal ganglia tumors

Ref. 7 - p. 346

SECTION IV - CEREBELLUM, DIENCEPHALON, RHINENCEPHALON, CEREBRUM

1321. THE LAMINATIONS OF THE CEREBRAL CORTEX ARE IN PART FORMED BY THE:
1. Molecular layer
2. Layer of pyramidal cells
3. Layer of fusiform cells
4. Ganglionic layer Ref. 7 - p. 351

1322. CONCERNING THE VISUAL RECEPTIVE CENTERS OF THE CEREBRUM:
1. Located in the cortex forming the walls of the calcarine fissure
2. Does not differ structurally from other cortical centers
3. Visual cortex is relatively thick
4. Lower half of area 17 receives visual impressions from objects in upper visual field Ref. 7 - pp. 369-370

1323. ASSOCIATED WITH THE TEMPORAL LOBE:
1. Olfactory impulses
2. Auditory receptive impulses
3. Vestibular receptive impulses
4. Vision Ref. 7 - p. 371

1324. ANATOMICALLY ASSOCIATED WITH THE FORNIX:
1. Corpus callosum
2. Septum pallucidum
3. Transverse fissure of cerebrum
4. Thalamus Ref. 7 - p. 337

1325. CONCERNING THE ANTERIOR COMMISSURE:
1. Belongs to the mesencephalon
2. Joins the optic ganglia
3. Found only in humans
4. Function unknown Ref. 7 - pp. 337-338

SECTION IV - CEREBELLUM, DIENCEPHALON, RHINENCEPHALON, CEREBRUM

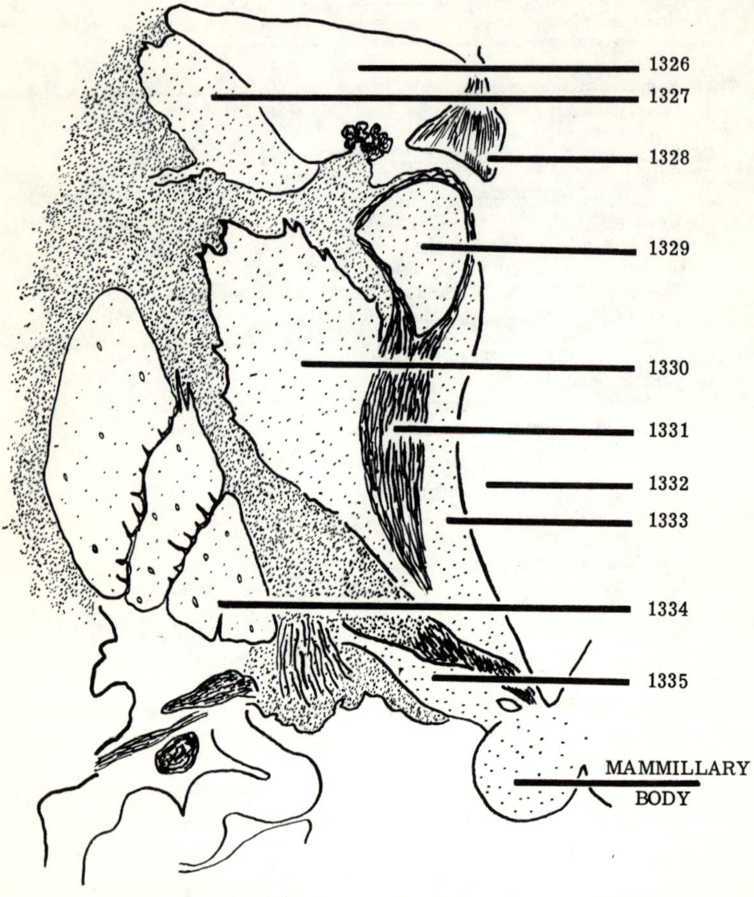

IDENTIFY THE STRUCTURES INDICATED BY THE GUIDE LINES IN A SECTION THROUGH THE THALAMIC REGION AT THE LEVEL OF THE MAMMILLARY BODIES:

A. Caudate nucleus
B. Fornix
C. Lateral part of thalamus
D. Putamen
E. Fasciculus lenticularis
F. Mammillothalamic tract
G. Anterior part of thalamus
H. Globus pallidus
I. Massa intermedia
J. Corona radiata
K. Hippocampus
L. Claustrum
M. Third ventricle
N. Anterior central gyrus
O. Lateral ventricle

Ref. 7 - pp. 521-522

SECTION V - PATHWAYS, CASE HISTORIES

ARRANGE THE GROUP OF FIVE ITEMS INTO A PROPER SEQUENCE:

1336. RECEPTION OF AN AUDITORY STIMULUS AND TRANSMISSION TO TRANSVERSE TEMPORAL GYRUS:
A. Inferior colliculus
B. Decussation
C. Nucleus of lateral lemniscus
D. Medial geniculate body
E. Dorsal cochlear nucleus Ref. 7 - p. 260

1337. TRANSMISSION OF VISUAL IMPULSES:
A. Optic chiasma
B. Layer of rods and cones
C. Ganglionic layer
D. Optic tract
E. Lateral geniculate body Ref. 7 - p. 265

1338. LAYERS OF THE CEREBRAL CORTEX, STARTING WITH THE MOST SUPERFICIAL:
A. Molecular layer
B. External granular layer
C. Layer of pyramidal cells
D. Internal granular layer
E. Ganglionic layer Ref. 7 - p. 351

1339. TACTILE PATHWAY:
A. Posterolateral ventral nucleus of thalamus
B. Medial lemniscus
C. Ventral spinothalamic tract
D. Mesencephalon
E. Internal capsule Ref. 7 - p. 384

1340. PAIN PATHWAY:
A. Spinal ganglion
B. Lateral spinothalamic tract
C. Internal capsule
D. Posterolateral ventral nucleus of thalamus
E. Decussation Ref. 7 - p. 387

1341. EXTEROCEPTIVE PATHWAY ASSOCIATED WITH TRIGEMINAL NERVE:
A. Medial lemniscus
B. Ventral secondary tract n. V
C. Decussation
D. Main sensory nucleus of V
E. Semilunar ganglion Ref. 7 - p. 389

1342. SPINAL PROPRIOCEPTIVE PATHWAY TO CEREBRAL CORTEX:
A. Medial lemniscus
B. Posterior funiculus of spinal cord
C. Posterior limb of internal capsule
D. Decussation
E. Cuneate nuclei of medulla oblongata Ref. 6 - p. 392

SECTION V - PATHWAYS, CASE HISTORIES

1343. CEREBELLAR CONNECTIONS OF THE VESTIBULAR NERVE:
A. Vestibular nerve
B. Vestibular nuclei
C. Vestibular ganglion
D. Cortex of the vermis
E. Vestibulocerebellar tract Ref. 6 - p. 393

1344. MOTOR PATHWAY FOR THE SPINAL NERVES:
A. Basis pedunculi of mesencephalon
B. Posterior limb of internal capsule
C. Partial decussation
D. Pyramidal cells
E. Pyramid of medulla oblongata Ref. 6 - p. 397

1345. CEREBELLORUBROSPINAL PATHWAY:
A. Decussation
B. Purkinje cells
C. Superior cerebellar peduncle
D. Nucleus ventralis lateralis of thalamus
E. Dentate nucleus Ref. 6 - p. 404

1346. PATHWAY FOR PRESSURE IMPULSES:
A. Spinal nerve
B. Decussation
C. Cell body in posterior gray column
D. Ventral spinothalamic tract
E. Posterior funiculus Ref. 6 - p. 385

1347. PATHWAY FOR TEMPERATURE SENSATIONS:
A. Gray matter of posterior spinal gray column
B. Lateral spinothalamic tract
C. Dorsal root
D. Decussation
E. Posterolateral ventral nucleus of thalamus
Ref. 6 - p. 386

1348. SPINAL PROPRIOCEPTIVE PATHWAY TO CEREBELLUM:
A. Ventral spinocerebellar tract
B. Intermediate gray matter of spinal cord
C. Anterior medullary velum
D. Bend around brachium conjunctivum
E. Rostral part of vermis Ref. 6 - p. 393

1349. MOTOR PATHWAY FOR CRANIAL NERVES:
A. Internuncial neurons
B. Genu of internal capsule
C. Striated musculature
D. Giant pyramidal cells
E. Large multipolar cells Ref. 7 - p. 399

SECTION V - PATHWAYS, CASE HISTORIES

1350. CORTICOPONTOCEREBELLAR PATHWAY:
A. Pyramidal cells
B. Anterior limb of internal capsule
C. Nuclei pontis
D. Decussation
E. Middle cerebellar peduncle Ref. 7 - p. 402

IN EACH OF THE FOLLOWING PATHWAYS THERE IS ONE COMPONENT WHICH IS INCORRECT. IDENTIFY THIS INCORRECT COMPONENT:

1351. VESTIBULAR REFLEX ARC THROUGH MEDIAL LONGITUDINAL BUNDLE:
A. Vestibular nuclei
B. Medial longitudinal fasciculus
C. Lateral lemniscus
D. Abducens nerve
E. Bipolar cells of vestibular ganglion Ref. 7 - p. 406

1352. PUPILLARY REFLEX:
A. Oculomotor nucleus
B. Retina
C. Partial decussation in optic chiasma
D. Posterior commissure
E. Synapses in pretectal region Ref. 7 - p. 409

1353. SPINAL PATHWAYS FOR SENSATIONS OF TOUCH:
A. Posterior funiculus
B. Unipolar neurons
C. Nucleus cuneatus
D. Nucleus gracilis
E. Medial lemniscus Ref. 7 - p. 385

1354 TEMPERATURE PATHWAY:
A. Spinal ganglion
B. Intercalated neurons
C. Lateral spinothalamic tract
D. Nucleus cuneatus
E. Mesencephalon Ref. 7 - p. 386

1355. PATELLAR TENDON REFLEX:
A. Golgi tendon organ
B. Dorsal root ganglion
C. Motor end plate
D. Alpha fiber
E. L-1 spinal segment Ref. 8 - p. 257

1356. TACTILE SENSE PATHWAY AND KINESTHETIC SENSE:
A. Pacinian corpuscle
B. Cerebellum
C. Pons
D. Posterior limb of internal capsule
E. Nucleus gracilis and cuneatus Ref. 8 - p. 260

SECTION V - PATHWAYS, CASE HISTORIES

1357. PAIN PATHWAY:
A. Meissner's corpuscle
B. Dorsal root ganglion
C. Pons
D. Lateral spinothalamic tract
E. Posterior limb of internal capsule Ref. 8 - p. 267

1358. VOLUNTARY MOTOR ACTIVITY (SKILLED ACT) PATHWAY:
A. Pyramidal decussation
B. Pons
C. Corticospinal tract
D. Precentral gyrus of cerebral cortex
E. Cerebellum Ref. 8 - p. 276

1359. AUDITORY PATHWAY:
A. Inferior colliculus of midbrain
B. Medial lemniscus
C. Medial geniculate body
D. Dorsal cochlear nucleus
E. Superior olivary nucleus Ref. 8 - p. 344

1360. AUDITORY PATHWAY:
A. Nucleus of trapezoid body
B. Nucleus of lateral lemniscus
C. Brachium of inferior colliculus
D. Pyramid of medulla
E. Temporal cerebral cortex Ref. 8 - p. 344

FOLLOWING THE CASE HISTORY ARE SEVERAL QUESTIONS BASED ON THE HISTORY. CHOOSE THE BEST ANSWER FOR EACH QUESTION.

A 42 year-old bartender presents with a cigarette burn of the right forefinger and thumb. The patient states that he was not aware of being burned and that he now notices that he feels no pain or temperature sensations in these fingers. On examination it is found that the patient indeed has a symmetrical deficiency involving the same fingers of both hands. However, sensations of touch and proprioception are not impaired. In addition, motor functions of these fingers are found to be normal.

1361. THE LESION IS LOCALIZED TO THE SPINAL CORD AT LEVELS:
A. C-2, C-3, C-4
B. C-5, C-6, C-7
C. C-8, T-1, T-2
D. T-3, T-4, T-5
E. T-6, T-7, T-8 Ref. 7 - p. 131

1362. TO INVOLVE BOTH EXTREMITIES THE LESION IS PROBABLY IN THE:
A. Dorsal root ganglion
B. Spinal nerve
C. Anterior white commissure
D. Lateral spinothalamic tract
E. Internal capsule Ref. 7 - pp. 168, 387, 435

SECTION V - PATHWAYS, CASE HISTORIES

1363. THE FIBERS INVOLVED ARE THOSE OF THE:
A. Spinothalamic tract
B. Lateral lemniscus
C. Medial lemniscus
D. Fasciculus gracilis
E. Rubrospinal tract Ref. 7 - p. 435

1364. SENSATION OF TOUCH IS NOT LOST BECAUSE:
A. Touch fibers are far removed from this lesion
B. Tactile fibers are found along autonomic pathways
C. Tactile fibers are intact at other levels and impulses get past the lesion
D. This is a most unusual occurrence
E. None of the above Ref. 7 - p. 189

1365. THIS DISEASE OF THE SPINAL CORD IS CALLED:
A. Brown-Sequard syndrome
B. Lateral sclerosis
C. Spina bifida occulta
D. Syringomyelia
E. Tabes dorsalis Ref. 7 - p. 435

A 38 year-old housewife gives a history of recurrent headaches of increasing severity, associated with visual disturbances. She reports "strange" attacks beginning with sensations of unpleasant odors. Recently the attacks have been more severe and she claims to "smell burning rubber". At times the patient has been noted to stare off into space, but no loss of consciousness has ever occurred. During her last attack, the patient had a visual hallucination of "a bright green tiger with a long red tail, standing on a brilliant blue tree stump". Neurological examination demonstrated a right homonymous hemianopsia, weakness of the right muscles of mastication, papilledema and no sensory deficits.

1366. A DIAGNOSIS IS MADE OF A TUMOR, MOST LIKELY INVOLVING THE:
A. Frontal lobe
B. Parietal lobe
C. Temporal lobe
D. Occipital lobe
E. Cerebellum Ref. 7 - pp. 269, 346

1367. THE IRRITATING LESION PRODUCING THE ODOR HALLUCINATIONS INVOLVES THE:
A. Uncus
B. Fornix
C. Olfactory bulb
D. Striae terminalis
E. Anterior medullary velum Ref. 7 - p. 346

SECTION V - PATHWAYS, CASE HISTORIES

1368. THE RIGHT HOMONYMOUS HEMIANOPSIA IN THIS CASE IS DUE TO INTERRUPTION OF THE:
A. Right geniculocalcarine fasciculus
B. Left geniculocalcarine fasciculus
C. Right occipital lobe
D. Left occipital lobe
E. Left optic nerve
Ref. 7 - pp. 267, 430

1369. THE RIGHT FACIAL WEAKNESS IS PROBABLY SECONDARY TO PRESSURE ON THE:
A. Hippocampus
B. Anterior central gyrus
C. Posterior central gyrus
D. Trigeminal nerve
E. Buccinator nerve
Ref. 7 - pp. 57, 358

1370. THE TUMOR IS LOCALIZED TO THE:
A. Left side
B. Right side
C. Mid-line
D. No conclusions can be made
E. Only pneumoencephalogram can reveal the location
Ref. 7 - p. 430

A 68 year-old woman complains of headaches and vertigo. Gradually the patient develops a spastic hemiplegia on the right side, with hyperactive deep tendon reflexes of the right arm and right leg. The patient complains of "double vision" and examination reveals that the left eye is deviated nasally, indicating a paralysis of the left external rectus muscle. No other ocular abnormalities are detected. The pupillary reactions are normal.

1371. THE LESION IS MOST PROBABLY LOCATED IN THE:
A. Medulla oblongata
B. Cerebellum
C. Pons
D. Mesencephalon
E. Cerebral cortex
Ref. 7 - p. 441

1372. THE LESION IS ON THE:
A. Right side
B. Left side
C. Mid-line
D. No conclusions can be reached
E. Bilateral
Ref. 7 - p. 441

1373. THE NERVE INVOLVED IS THE:
A. Optic
B. Oculomotor
C. Trochlear
D. Abducens
E. Trigeminal
Ref. 7 - p. 441

SECTION V - PATHWAYS, CASE HISTORIES

A 53 year-old bank teller complains of "increasing stiffness" in the legs. He is advised to try a new pair of rubber sole shoes since his job requires long periods of standing. Gradually the patient is unable to lift his feet from the ground. On examination the leg appeared to be moved as one unit from the hip down. The deep tendon reflexes were hyperactive. No muscle atrophy or sensory losses could be found. The patient in general appeared to have lost the voluntary control of the leg muscles. Hematological survey revealed no abnormalities; urine analysis was within normal limits.

1374. THE MOST LIKELY DIAGNOSIS IS:
A. Poliomyelitis
B. Tabes dorsalis
C. Brown-Sequard syndrome
D. Lateral sclerosis
E. Hysterical reaction — Ref. 7 - pp. 433-434

1375. THIS CONDITION DIFFERS FROM SYRINGOMYELIA BECAUSE OF:
A. Changes in deep tendon reflexes
B. Disturbances of motor functions
C. Lack of muscle atrophies
D. Lack of sensory disturbances
E. All of the above — Ref. 7 - pp. 434-435

1376. THE TRACTS APPARENTLY INVOLVED IN THIS CASE ARE THE:
A. Corticospinal
B. Corticothalamic
C. Corticocerebellar
D. Rubrospinal
E. Spinothalamic — Ref. 7 - pp. 397-399

A 59 year-old retired army major suddenly falls to the ground while raking leaves in his yard. He is unconscious for three hours, following which he slowly recovers. The patient, following appropriate diagnostic examinations, is found to have had a cerebrovascular accident. The lesion has involved the left side of the medulla oblongata with more specific involvement of the pyramid and the emerging fibers of the hypoglossal nerve. Neurological examination performed several months after the initial attack, reveals several sequelae.

1377. WHEN THE TONGUE IS PROTRUDED:
A. It deviates to the right
B. It deviates to the left
C. It deviates cranially
D. It deviates caudally
E. It rolls up — Ref. 7 - p. 438

1378. THE LEFT EYE AT REST:
A. Deviates to the right
B. Deviates to the left
C. Deviates upward
D. Deviates downward
E. Should not deviate — Ref. 7 - pp. 437-438

SECTION V - PATHWAYS, CASE HISTORIES

1379. THE LEFT ARM AND LEG REVEAL:
A. Spastic paralysis
B. Flaccid paralysis
C. Weakness but no paralysis
D. Tremors
E. No abnormalities Ref. 7 - p. 437

1380. THE RIGHT ARM AND LEG REVEAL:
A. Spastic paralysis
B. Flaccid paralysis
C. Weakness but no paralysis
D. Tremors
E. No abnormalities Ref. 7 - p. 437

A 32 year-old shoe salesman presents with a two-month history of pain in the abdomen. The pain is described as being a "strip on the right side from the navel to the back". One month after onset of this symptom, he noted weakness in the right leg. Neurological examination revealed a loss of pain, temperature and perception sensations on the right side, from the level of the umbilicus down. A spastic paralysis of the right lower extremity is found. Appropriate studies indicate that the patient has a tumor of the dorsal root pressing on the spinal cord.

1381. THE TUMOR IS LOCATED AT LEVEL:
A. T-2
B. T-4
C. T-6
D. T-8
E. T-10 Ref. 7 - p. 130

1382. THE LESION IS:
A. Mid-line
B. On the right side
C. On the left side
D. Bilateral
E. Cannot be determined Ref. 7 - pp. 193-94, 198-99

A 60 year-old woman has been treated by her family physician for hypertension for six years. While working in the kitchen, she suddenly falls to the floor and remains unconscious for two hours. She is diagnosed as having had a cerebrovascular accident. The patient is found to have impairment of sensation of the entire left side of her body. Minor pin-prick stimulation on the left elicits marked emotional reactions. The patient describes the sensation as "having a red-hot poker jabbed into my skin".

1383. THE MOST LIKELY DIAGNOSIS:
A. Hysterical reaction
B. Dorsal root involvement
C. Damage to cerebral cortex
D. Involvement of thalamus
E. Schizophrenia Ref. 7 - p. 299

SECTION V - PATHWAYS, CASE HISTORIES

1384. IF THE LESION IS IN THE THALAMUS, THE AREA INVOLVED IS THE:
A. Lateral nucleus
B. Anterior thalamic nucleus
C. Nuclei of the mid-line
D. Medial nuclei
E. None of the above Ref. 7 - p. 299

1385. THE LESION IS LOCATED:
A. In the mid-line
B. On the right side
C. On the left side
D. Bilateral
E. Cannot be determined Ref. 7 - p. 299

FOR EACH OF THE FOLLOWING MULTIPLE CHOICE QUESTIONS SELECT THE ONE MOST APPROPRIATE ANSWER.

1386. A DEVELOPMENTAL INCREASE IN MASS IS CALLED:
A. Hisitogenesis
B. Morphogenesis
C. Growth
D. Proliferation
E. Fertilization Ref. 9 - p. 14

1387. DIVERSITY OF FORM IS BROUGHT ABOUT BY:
A. Differential rates of growth
B. Metaplasia
C. Absolute growth
D. Protoplasmic synthesis
E. None of the above Ref. 9 - p. 15

1388. AT BIRTH THE SURFACE AREA OF THE NEWBORN AVERAGES ____________ SQUARE CM.:
A. 500
B. 1000
C. 2000
D. 3000
E. 4000 Ref. 9 - p. 17

1389. FACTORS CONTROLLING GROWTH:
A. Temperature
B. Hormones
C. Nutritional factors
D. All of the above
E. None of the above Ref. 9 - pp. 18-19

1390. ACHONDROPLASIA IS CHARACTERIZED BY ALL OF THE FOLLOWING, EXCEPT:
A. Inferior sexual development
B. Short limbs
C. Average intelligence
D. Broad, flat face
E. Undershot jaw Ref. 9 - p. 22

1391. SPECIFIC MORPHOGENETIC PROCESSES INCLUDE:
A. Cell migration
B. Fusion and splitting
C. Folding
D. All of the above
E. None of the above Ref. 9 - p. 24

1392. GENES IMPRESS A MESSAGE ON "MESSENGER ____________":
A. ATP
B. DNA
C. RNA
D. ADP
E. None of these Ref. 9 - p. 25

1393. THE HUMAN MALE CELL HAS __________ CHROMOSOMES:
A. 48
B. 46
C. 47
D. 24
E. 26 Ref. 9 - p. 31

1394. AS THE FOLLICLE GROWS, THE OÖGONIUM BECOMES LOCATED IN A MOUND OF FOLLICLE CELLS CALLED THE:
A. Vesicular antrum
B. Oöcyte
C. Cumulus oöphorus
D. Liquor folliculi
E. None of these Ref. 9 - p. 33

1395. THERE ARE APPROXIMATELY __________ MILLION SPERM IN THE AVERAGE EJACULATE OF 3.5 CC:
A. 5
B. 50
C. 100
D. 200
E. More than 200 Ref. 9 - p. 53

1396. THE MEETING AND UNION OF HUMAN SEX CELLS IS BELIEVED TO OCCUR IN THE:
A. Upper third of the oviduct
B. Middle third of the oviduct
C. Lower third of the oviduct
D. Uterus
E. Surface of the vagina Ref. 9 - p. 58

1397. SEXUALITY IS ESTABLISHED AT THE TIME OF:
A. Ovaluation
B. Gametogenesis
C. Histogenesis
D. Morphogenesis
E. None of the above Ref. 9 - p. 59

1398. CLEAVAGE DIVISIONS ARE ALWAYS:
A. Meiotic
B. Mitotic
C. Amitotic
D. Anucleotic
E. By binary fission Ref. 9 - pp. 63-64

1399. DERIVED FROM ENTODERM:
A. Muscle
B. Kidney
C. Gonads
D. Tonsils
E. Hair Ref. 9 - p. 83

1400. TISSUES DERIVED FROM MESODERM AND ECTODERM INCLUDE:
A. Epithelium of urethra
B. Epithelium of larynx
C. Epithelium of ureter
D. Epithelium of bladder
E. Epithelium of stomach Ref. 9 - p. 83

1401. DERIVED FROM ENTODERM:
A. Hypophysis
B. Epithelium of pharynx
C. Epithelium of anal canal
D. Enamel of teeth
E. Muscle Ref. 9 - p. 83

1402. THE DEFINITIVE YOLK SAC OF THE EMBRYO APPEARS BY THE ____________ DAY:
A. 13th
B. 26th
C. 72nd
D. 94th
E. 2nd Ref. 9 - p. 87

1403. AT THE CAUDAL END OF THE PRIMITIVE STREAK, ECTODERM AND ENTODERM FUSE AS THE:
A. Notochordal canal
B. Coelom
C. Cloacal membrane
D. Neural groove
E. Notochord Ref. 9 - p. 91

1404. THE NOTOCHORD IS REPLACED BY THE:
A. Ependyma
B. Vertebral column
C. Spinal canal
D. Dorsal roots
E. Spinal cord Ref. 9 - p. 93

1405. THE SPACE BETWEEN THE SPLIT LAYERS OF MESODERM IS THE:
A. Amniotic cavity
B. Blastocele
C. Foregut
D. Hindgut
E. None of the above Ref. 9 - p. 94

1406. DURING THE PERIOD WHEN THE BODY AND ITS PARTS ARE BEING LAID DOWN, DEVELOPMENT AND DIFFERENTIATION APPEAR FIRST IN THE ____________ REGION:
A. Head
B. Tail
C. Mid-embryo
D. Distal third
E. Mid-third Ref. 9 - p. 96

1407. IN THE FOURTH WEEK OF EMBRYONIC LIFE THERE ARE ABOUT ____________ SOMITES:
A. 6
B. 8
C. 10
D. 12
E. Less than 4 Ref. 9 - p. 97

1408. THE EMBRYO IS CALLED A FETUS AFTER THE:
A. Third week
B. Second month
C. Third month
D. Fourth month
E. Sixth month Ref. 9 - p. 98

1409. LANUGO COVERS THE BODY, SOME HEAD HAIRS SHOW AND FETAL MOVEMENTS ARE FELT BY THE MOTHER WHEN THE FETUS IS ABOUT ____________ MONTHS:
A. 5
B. 6
C. 7
D. 8
E. 9 Ref. 9 - p. 100

1410. ALMOST ALL OF THE INTERNAL ORGANS ARE WELL LAID DOWN AT ____________ MONTHS:
A. 1
B. 2
C. 3
D. 4
E. 5 Ref. 9 - p. 100

1411. THE TESTES BEGIN TO DESCEND INTO THE SCROTUM AT ABOUT A FETAL AGE OF ____________ MONTHS:
A. 4
B. 5
C. 6
D. 7
E. 8 Ref. 9 - p. 100

1412. A "STONE CHILD", A CALCIFIED DEAD FETUS, IS MORE PROPERLY CALLED A:
A. Placenta Praevia
B. Cyclops
C. Lithopedion
D. Fetus papyraceus
E. Abortion Ref. 9 - p. 102

1413. IT IS APPROXIMATELY CORRECT TO COMPUTE THE AGE OF AN EMBRYO FROM THE FOURTEENTH DAY AFTER:
A. Onset of last menstruation
B. Last sexual intercourse
C. Last full moon
D. Last missed menstrual period
E. Onset of breast changes Ref. 9 - p. 103

1414. THE YOLK SAC DETACHES FROM THE GUT BY THE END OF THE:
A. 2nd week
B. 5th week
C. 3rd month
D. 4th month
E. 5th month Ref. 9 - p. 107

1415. A PERSISTENCE OF THE PROXIMAL END OF THE YOLK SAC IN AN ADULT IS CALLED THE:
A. Urachal ligament
B. Appendix
C. Vitelline membrane
D. Ductus saccus
E. None of the above Ref. 9 - p. 109

1416. WHEN THE AMOUNT OF AMNIOTIC FLUID EXCEEDS TWO LITERS THE CONDITION IS CALLED:
A. Oliogohydramnios
B. Polyhydramnios
C. Amniotitis
D. Bag of waters
E. All of the above Ref. 9 - p. 110

1417. WHICH ONE OF THE FOLLOWING ANOMALIES IS MORE COMMON IN LIVE-BORN MALES:
A. Pyloric stenosis
B. Dislocated hip
C. Birthmarks
D. Brain defects
E. All of the above Ref. 9 - p. 185

1418. USUALLY ASSOCIATED WITH AN ABSENCE OF THE BRAIN:
A. Macrocephalus
B. Hydrocephalus
C. Cranioschisis
D. Microcephalus
E. Conical cranium Ref. 9 - p. 200

1419. IN THE PROCESS OF FACE FORMATION OR CONSTRUCTION, THE FIRST TO COME INTO BEING IS THE:
A. Orbital ridge
B. Upper jaw
C. Lower jaw
D. Ears
E. Nose Ref. 9 - p. 205

1420. AN ABSENT LOWER JAW IS CALLED:
A. Micrognathus
B. Cleft lip
C. Macrostomus
D. Cheiloschisis
E. Agnathus Ref. 9 - p. 206

1421. THE EMBRYONIC TAIL IS AT ITS MAXIMUM LENGTH ABOUT THE:
A. 5th week
B. 3rd month
C. 5th month
D. 7th month
E. 9th month Ref. 9 - p. 208

1422. DIGITS WHICH ARE ABNORMALLY SHORT ARE CALLED:
A. Syndactyly
B. Polydactyly
C. Dichuris
D. Amelus
E. None of the above Ref. 9 - p. 212

1423. THE FIRST INDICATION OF TOOTH DEVELOPMENT IS THE:
A. Dental bud
B. Bud state
C. Cap stage
D. Dental lamina
E. None of the above Ref. 9 - p. 217

1424. THE DECIDUOUS CANINE IS SHED AT ABOUT ____________ YEARS:
A. 6
B. 8
C. 10
D. 12
E. 16 Ref. 9 - p. 223

1425. THE PERMANENT CENTRAL INCISOR ERUPTS AT ABOUT ____________ YEARS:
A. 7
B. 9
C. 12
D. 14
E. 16 Ref. 9 - p. 224

ANSWER THE FOLLOWING QUESTIONS ACCORDING TO THE KEY BELOW:
A. If both statement and reason are true and related cause and effect
B. If both statement and reason are true but not related cause and effect
C. If the statement is true, but the reason is false
D. If the statement is false but the reason is true
E. If both statement and reason are false

1426. Physiologically the allantoic tube is the significant structure for maternal-fetal circulation

because

It is responsible for anomalies of the urachus.
Ref. 9 - pp. 110-111

1427. The human blastocyst probably begins uterine attachment ten days after ovulation

because

Shortly before implantation the encapsulating zona pellucida disappears. Ref. 9 - pp. 121-123

1428. Both the cellular and syncitial trophoblast ingest maternal tissue
because
Reticular fibers are common elements associated with young trophoblastic cells. Ref. 9 - pp. 124-128

1429. The mucosa of the pregnant uterus is named decidua
because
It falls off after delivery. Ref. 9 - pp. 128-131

1430. The placenta is of double origin
because
It is composed of fetal and maternal contributions.
Ref. 9 - pp. 131-135

1431. The chorionic villi are the most important parts of the placenta
because
It attaches the placenta to the endometrium. Ref. 9 - p. 133

1432. The placenta is a semi-permeable barrier
because
It permits the passage of bacteria and viruses to the embryo.
Ref. 9 - pp. 138-140

1433. The changes that characterize the female sexual cycle are not confined to the ovaries and uterus
because
Growth changes and regression take place in the mammary glands in coordination with the ovarian cycle. Ref. 9 - p. 155

1434. Estrogen is especially associated with the corpus luteum
because
It produces most of the characteristic features of the menstrual cycle. Ref. 9 - p. 158

1435. Cleavage is a mechanism determining particular parts of the embryo
because
Loss of cleavage cells (experimentally or naturally) results in specific fetal losses. Ref. 9 - pp. 170-171

1436. Congential anomalies are gross anatomical defects easily observed at birth
because
Acquired defects result from mechanical weakness or disease.
Ref. 9 - p. 184

1437. Certain human syndromes are associated with abnormal chromosome numbers
because
An individual with 48 chromosomes has mongolism.
Ref. 9 - p. 187

1438. Cranioschisis implies hydrocephalus
because
It is associated with an absence of the brain
Ref. 9 - p. 200

1439. The development of the face occurs between the fifth and eighth weeks
because
Progressive modelling of the face continues throughout childhood.
Ref. 9 - pp. 204-206

1440. Cleft lip is usually bilateral
because
It is associated with a failure of fusion of the second branchial arch.
Ref. 9 - p. 207

ANSWER THE FOLLOWING QUESTIONS ACCORDING TO THE KEY BELOW:
A. If only A is correct
B. If only B is correct
C. If A and B are correct
D. If A and B are incorrect

1441. DURING EMBRYOLOGICAL DEVELOPMENT, THE LARGE INTESTINE IS:
A. Last to leave the umbilical cord and re-enter the abdominal cavity
B. Always larger in length than the small intestine
C. Both
D. Neither Ref. 9 - pp. 251-253

1442. TYPICAL BILE, SECRETED BY HEPATIC CELLS:
A. Occurs in fetuses 5 months old
B. Colors the meconium
C. Both
D. Neither Ref. 9 - p. 257

1443. ACCESSORY PANCREASES:
A. Are extremely rare
B. Occur within the wall of the intestine and stomach
C. Both
D. Neither Ref. 9 - p. 262

1444. THE EPIPLOIC FORAMEN IS AN OPENING BETWEEN THE:
A. Vestibule of omental bursa
B. Peritoneal cavity
C. Both
D. Neither Ref. 9 - p. 277

1445. CONTRIBUTING TO THE FORMATION OF THE DIAPHRAGM:
A. Septum Transversum
B. Mesenchyme of the costal body walls
C. Both
D. Neither Ref. 9 - p. 293

1446. DERIVATIVES OF THE HUMAN PRONEPHROS:
A. Renal capsule
B. Bowman's capsule
C. Both
D. Neither Ref. 9 - p. 297

1447. FUNCTIONAL DERIVATIVES OF THE MESONEPHROS:
A. Efferent ductules of male
B. Fallopian tube of female
C. Both
D. Neither Ref. 9 - pp. 298-300

1448. PARTITIONING OF THE ATRIUM IS ACCOMPANIED BY GROWTH OF:
A. Septum primum
B. Septum secundum
C. Both
D. Neither Ref. 9 - pp. 381-382

1449. TWO OF THE FOUR FEATURES OF TETRALOGY OF FALLOT INCLUDE:
A. Interatrial septal defect
B. Stenosis of pulmonary artery
C. Both
D. Neither Ref. 9 - p. 391

1450. THE METENCEPHALON FORMS THE:
A. Cerebellum
B. Medulla oblongata
C. Both
D. Neither Ref. 9 - p. 481

1451. THE MESENCEPHALON DIFFERENTIATES INTO THE:
A. Cerebellum
B. Pons
C. Both
D. Neither Ref. 9 - p. 481

1452. AT BIRTH, THE DISTAL PORTIONS OF BOTH INTERNAL ILIAC ARTERIES COLLAPSE AND PERSIST AS THE:
A. Urachus
B. Lateral umbilical ligaments
C. Both
D. Neither Ref. 9 - p. 357

1453. FACTORS ASSISTING IN CAUSING ROTATION OF THE STOMACH:
A. Rapid expansion of the dorsal mesentery
B. Slow growth of the vertebral mesentery
C. Both
D. Neither Ref. 9 - p. 248

1454. THE OPTIC CUP IS DESTINED TO BECOME THE:
A. Ciliary body
B. Retina
C. Both
D. Neither Ref. 9 - p. 530

1455. AUDITORY OSSICLES DEVELOP FROM THE CONDENSED MESENCHYME OF THE:
A. Third branchial arch
B. Fourth branchial arch
C. Both
D. Neither Ref. 9 - p. 547

FOLLOWING EACH GROUP OF FIVE CHOICES THERE ARE EITHER FIVE STATEMENTS OR FIVE ITEMS. MATCH THE STATEMENTS OR ITEMS WITH THE MOST APPROPRIATE CHOICE GIVEN. EACH CHOICE MAY BE USED ONLY ONCE:

A. Bulbar ridges
B. Left sinal valve
C. Endocardial cushions
D. Left sinal horn
E. Truncus

1456. ___ Semilunar valves

1457. ___ Contributes to atrial septum

1458. ___ Oblique vein

1459. ___ Tricuspid valve

1460. ___ Pulmonary artery Ref. 9 - pp. 377-378

A. Mullerian duct
B. Genital swellings
C. Vesico-urethral primordium
D. Wolffian duct
E. Gonad

1461. ___ Posterior commissure

1462. ___ Upper prostatic urethra

1463. ___ Seminal vesicle

1464. ___ Rete testis

1465. ___ Appendix testis Ref. 9 - pp. 319, 325, 328

A. Yolk sac
B. Allantois
C. Amnion
D. Chorion
E. Decidua basalis

1466. ___ Frondose part becomes fetal placenta

1467. ___ Covers fetal surface of placenta and afterbirth

1468. ___ Between chorion and muscular wall of uterus

1469. ___ Roof forms the gut

1470. ___ First lies within the body stalk Ref. 9 - p. 137

A. Aphakia
B. Coloboma
C. Myeloschisis
D. Hyperthelia
E. Anonychia

1471. ___ Absence of the lens

1472. ___ Absence of a sector of the iris

1473. ___ Widely exposed spinal cord

1474. ___ Accessory nipples

1475. ___ Absence of nails

Ref. 9 - pp. 444,452,468, 540-541

A. Aprosopus
B. Agnathus
C. Cheiloschisis
D. Rachischisis
E. Uranoschisis

1476. ___ Cleft lip

1477. ___ Cleft palate

1478. ___ Featureless face

1479. ___ Absent lower jaw

1480. ___ Cleft spine (spina bifida)

Ref. 9 - pp. 206-207, 209,230

IDENTIFY THE COMPONENTS OF THE SCHEMATIC TRANSVERSE SECTION OF THE VERTEBRATE EMBRYO.

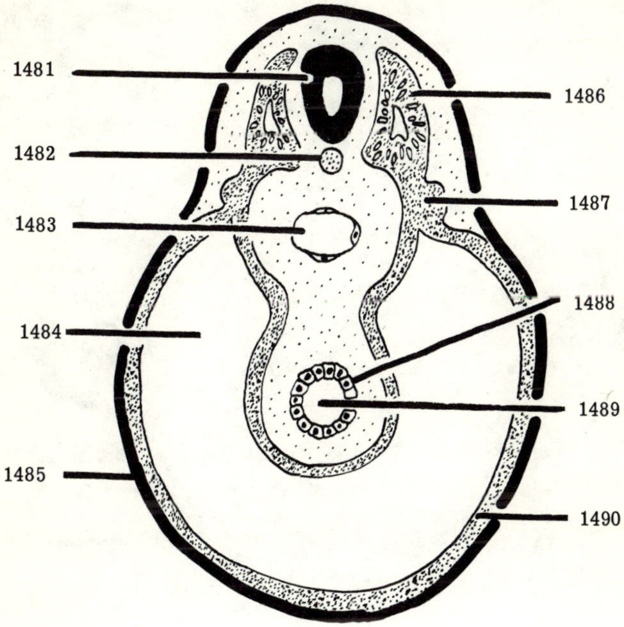

A. Aorta
B. Splanchnic mesoderm
C. Entoderm
D. Notochord
E. Neural tube
F. Gut
G. Somatic mesoderm
H. Myotome
I. Yolk stalk
J. Allantois
K. Chorion
L. Nephrotome
M. Coelom
N. Somite
O. Ectoderm

Ref. 9 - p. 93

IDENTIFY EACH OF THE FOLLOWING CONGENITAL ANOMALIES.

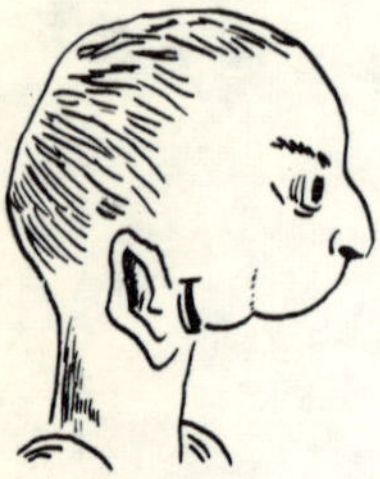

1491

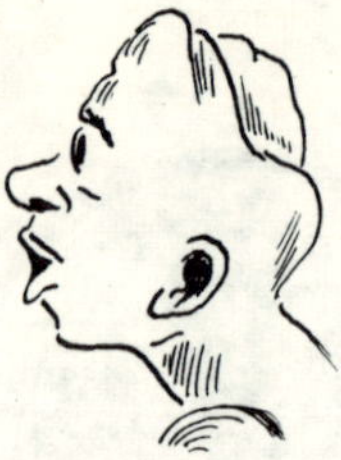

1492

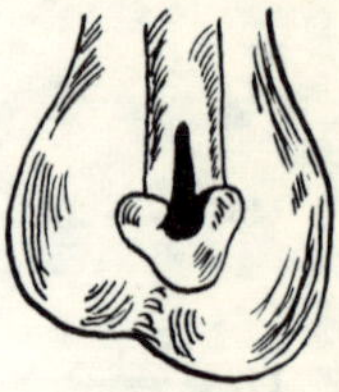

1493

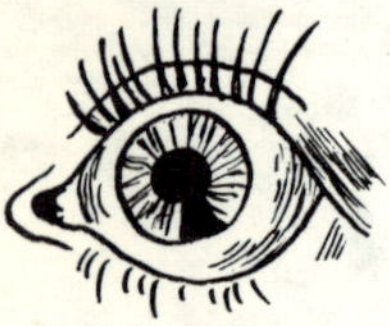

1494

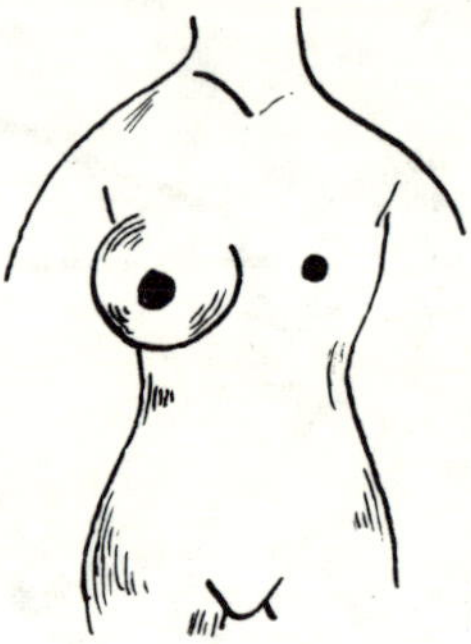

1495

1496

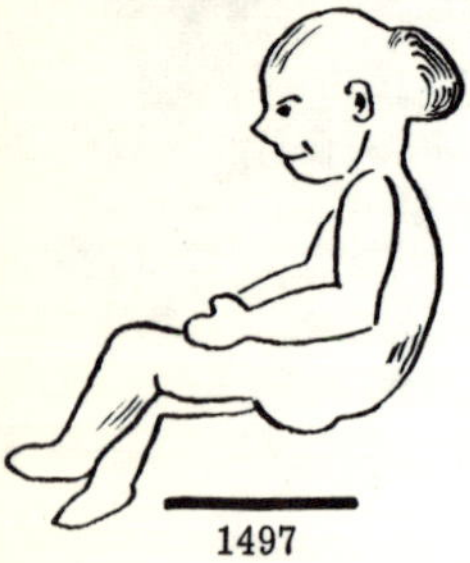

1497

1498

1499

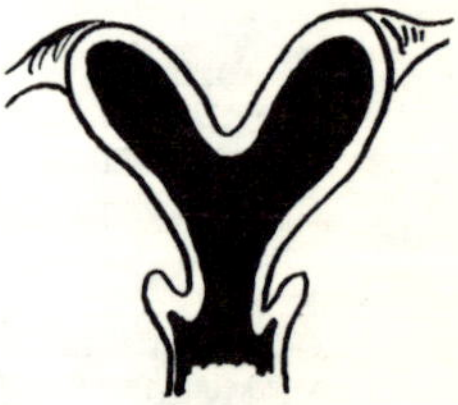

1500

A. Epispadias
B. Encephalocele
C. Cranioschisis
D. Cleft hand
E. Amelius
F. Coloboma
G. Hypospadias
H. Agnathus
I. Syndactyly
J. Bicornuate uterus
K. Hermaphrodite
L. Polydactyly
M. Amastia
N. Double uterus
O. Recto-urethral fistula

Ref. 9 - pp. 200. 206, 212, 313, 329, 338, 452, 498, 541

GROSS ANATOMY

1. "An Atlas of Anatomy"
J. C. B. Grant
5th Edition, 1962
Williams and Wilkins Company
Baltimore, Maryland

2. "Grant's Method of Anatomy"
J. C. B. Grant and J. V. Basmajian
7th Edition, 1965
Williams and Wilkins Company
Baltimore, Maryland

3. "Anatomy of the Human Body"
H. Gray
28th Edition, 1966
Lea and Febiger
Philadelphia, Pennsylvania

4. "Morris' Human Anatomy"
B. J. Anson
12th Edition, 1966
McGraw-Hill Book Co., Inc.
New York, New York

5. "Essentials of Human Anatomy"
R. T. Woodburne
4th Edition, 1969
Oxford University Press
London and Toronto

HISTOLOGY

6. "A Textbook of Histology"
W. Bloom and D. W. Fawcett
9th Edition, 1968
W. B. Saunders Company
Philadelphia, Pennsylvania

NEUROANATOMY

7. "The Anatomy of the Nervous System"
S. W. Ranson and S. L. Clark
10th Edition, 1959
W. B. Saunders Company
Philadelphia, Pennsylvania

8. "Human Neuroanatomy"
R. C. Truex and M. B. Carpenter
6th Edition, 1969
Williams and Wilkins Company
Baltimore, Maryland

EMBRYOLOGY AND TERATOLOGY

9. "Developmental Anatomy"
L. B. Arey
7th Edition, 1965
W. B. Saunders Company
Philadelphia, Pennsylvania

ANSWER KEY

The author has taken great pains to check thoroughly all questions and answers. However, in a volume of this size some inaccuracies and ambiguities may appear. Therefore, if in doubt, consult your references.

THE PUBLISHERS

GROSS ANATOMY

1. C
2. B
3. B
4. A
5. C
6. D
7. E
8. B
9. D
10. A
11. A
12. C
13. B
14. B
15. E
16. A
17. B
18. B
19. C
20. A
21. E
22. E
23. C
24. B
25. A
26. D
27. B
28. E
29. A
30. E
31. A
32. B
33. D
34. A
35. C
36. B
37. E
38. D
39. B
40. B
41. B
42. E
43. C
44. C
45. D
46. E
47. C
48. C
49. B
50. B
51. E
52. E
53. A
54. B
55. E
56. C
57. D
58. A
59. B
60. C
61. E
62. C
63. A
64. C
65. A
66. D
67. E
68. C
69. D
70. A
71. B
72. A
73. D
74. D
75. A
76. D
77. B
78. E
79. B
80. B
81. C
82. D
83. B
84. E
85. B
86. B
87. D
88. A
89. E
90. C
91. C
92. A
93. E
94. C
95. E
96. E
97. C
98. D
99. A
100. B
101. B
102. A
103. E
104. C
105. E
106. C
107. A
108. A
109. E
110. C
111. D
112. E
113. D
114. E
115. A
116. B
117. C
118. E
119. E
120. A
121. A
122. B
123. C
124. A
125. D
126. E
127. B
128. D
129. E
130. B
131. A
132. C
133. D
134. E
135. B
136. B
137. A
138. C
139. E
140. D
141. B
142. A
143. C
144. D
145. E
146. D
147. C
148. B
149. E
150. A
151. C
152. B
153. A
154. E
155. D
156. D
157. E
158. A
159. B
160. C
161. E
162. A
163. N
164. C
165. O
166. L
167. J
168. I
169. B
170. M
171. B
172. D
173. K
174. N
175. F
176. I
177. A
178. B
179. N
180. M
181. A
182. B
183. C
184. D
185. E
186. F
187. G
188. H
189. I
190. J
191. E
192. M
193. C
194. O
195. L
196. I
197. F
198. K
199. D
200. B
201. C
202. A
203. L
204. B
205. M
206. O
207. E
208. F
209. J
210. I
211. A
212. B
213. C
214. H
215. I
216. J
217. M
218. N
219. O
220. D
221. E
222. H
223. A
224. G
225. B
226. F
227. I
228. C
229. J
230. D

ANSWER KEY

GROSS ANATOMY

231. A, C
232. A, B, C
233. B
234. D, E
235. A, B, C, D
236. A, E
237. A, B
238. E
239. A, B, C, D, E
240. A, D
241. B, E
242. D
243. A
244. A, B, C, D
245. C, D
246. A, B, C, D, E
247. A, B, C, D, E
248. A, B, C, D
249. C
250. A, D
251. A, B, C
252. A, C, E
253. C, D, E
254. E
255. A, B
256. B, D, E
257. A, B, C
258. A, B, C, D, E
259. B, C, D
260. A, B
261. A, B, C
262. A, C, E
263. D
264. A, B
265. A
266. A, B, D, E
267. C, E
268. D
269. B, C, D
270. A
271. A
272. C, D, E
273. B
274. B, C
275. A, B, C, D, E
276. A, B
277. C
278. B
279. A, B, D
280. C, D
281. C
282. D
283. A
284. C
285. B
286. D
287. D
288. C
289. B
290. D
291. D
292. B
293. A
294. C
295. C
296. D
297. B
298. A
299. C
300. D
301. D
302. A
303. F
304. E
305. B
306. D
307. C
308. C
309. A
310. D
311. E
312. A
313. D
314. F
315. A
316. B
317. E
318. F
319. E
320. A
321. C
322. B
323. D
324. A
325. B
326. A
327. D
328. E
329. D
330. E
331. E
332. E
333. C
334. C
335. E
336. B
337. E
338. A
339. D
340. C
341. C
342. B
343. D
344. E
345. A
346. D
347. E
348. A
349. B
350. C
351. O
352. L
353. A
354. B
355. N
356. J
357. K
358. C
359. G
360. E
361. E
362. I
363. A
364. B
365. J
366. G
367. A
368. F
369. C
370. H
371. F
372. E
373. A
374. M
375. O
376. G
377. I
378. D
379. H
380. B
381. D
382. D
383. B
384. C
385. A
386. A
387. C
388. E
389. D

GROSS ANATOMY

390. C
391. C
392. B
393. D
394. B
395. A
396. E
397. C
398. D
399. C
400. E
401. A
402. D
403. C
404. B
405. D
406. B
407. C
408. D
409. B
410. D
411. A
412. E
413. A
414. E
415. B
416. C
417. A
418. A
419. E
420. B
421. A
422. E
423. C
424. D
425. A
426. C
427. C
428. A
429. E
430. E
431. A
432. B
433. E
434. C
435. B
436. B
437. C
438. C
439. B
440. A
441. E
442. D
443. B
444. A
445. E
446. B
447. C
448. E
449. E
450. C
451. B
452. C
453. B
454. A
455. D
456. A
457. B
458. A
459. A
460. C
461. C
462. A
463. B
464. A
465. D
466. D
467. B
468. C
469. C
470. D
471. A
472. D
473. A
474. C
475. C
476. C
477. A
478. D
479. C
480. A
481. D
482. A
483. C
484. C
485. B
486. D
487. E
488. E
489. F
490. A
491. K
492. A
493. M
494. B
495. O
496. E
497. N
498. G
499. I
500. E
501. C
502. M
503. B
504. F
505. L
506. K
507. N
508. J
509. E
510. O
511. A
512. D
513. F
514. B
515. C
516. N
517. J
518. K
519. I
520. O
521. A, B, C, E
522. B, C, E
523. A, B, C, D, E
524. A, E
525. E
526. A
527. A, B, C, D
528. B, C, D
529. A, E
530. A, B, C, D
531. A, B, C, D, E
532. A, B, C, D, E
533. B
534. A, E
535. A
536. A, B, C
537. A, E
538. A, E
539. B, D
540. B
541. B, C, D, E
542. A, C, E
543. B, C
544. A, B, E
545. A, B, C, D
546. A, B, C
547. A, B, C, D
548. A, B, E

GROSS ANATOMY

549. B, C
550. C, D
551. A, B, C
552. A, B, C, D
553. A, B
554. C
555. A, B, C, D
556. C, D, E
557. A, B, C, D, E
558. A, B, E
559. A, C, D, E
560. A, B, C, D, E
561. E
562. D
563. D
564. A
565. B
566. D
567. F
568. D
569. C
570. A
571. E
572. D
573. A
574. B
575. E
576. C
577. F
578. D
579. A
580. C
581. D
582. B
583. D
584. C
585. E
586. C
587. A
588. B
589. E
590. A
591. J
592. O
593. E
594. C
595. M
596. K
597. A
598. N
599. G
600. B
601. E
602. C
603. M
604. O
605. A
606. F
607. N
608. I
609. K
610. B
611. G
612. L
613. A
614. O
615. O
616. G
617. B
618. A
619. E
620. C
621. A, B, C, D, E
622. C, D, E
623. A, B, C, E
624. B, C, D, E
625. A
626. A, D, E
627. A, B, C, D, E
628. A, C, E
629. A, B, D, E
630. A, E
631. A, B
632. D, E
633. A, B, C
634. A, C, E
635. D
636. C
637. B, D, E
638. A, B, D
639. A, E
640. A, D
641. E
642. C, D
643. D
644. A, B, C
645. A, B, C, D
646. A, E
647. A, B, C
648. D, E
649. A, B, C, E
650. A, B, E
651. D
652. A, D, E
653. A, B, C, D, E
654. C, D, E
655. B, C
656. A, B, C, D
657. A, B, C
658. A, B, C
659. A, C, D
660. E
661. A, B
662. A, D
663. A, B, C, D
664. C, D
665. A, B, C, D
666. E
667. C, D, E
668. A, B, C, D
669. A, B
670. D
671. A, C, E
672. A, C, E
673. B, E
674. A, B, C, D, E
675. C
676. B, E
677. C, D, E
678. A, B, C
679. A, B, C
680. B, C, D
681. C, D
682. A, E
683. A, B, C
684. A, C, E
685. B, C
686. B
687. C
688. D
689. B
690. E
691. A
692. B
693. A
694. E
695. A
696. B
697. A
698. C
699. A
700. D
701. E
702. E
703. E
704. E
705. D
706. A
707. C

GROSS ANATOMY

708. A
709. B
710. D
711. C
712. E
713. C
714. B
715. D
716. A-2
717. B-1
718. E-1
719. C-3
720. A-1
721. C-1
722. D-2
723. D-3
724. A-1
725. B-3
726. D
727. E
728. B
729. D
730. A
731. F
732. L
733. A
734. H
735. C
736. J
737. D
738. C
739. B
740. O
741. A
742. B
743. C
744. D
745. E
746. F
747. G
748. H
749. I
750. J
751. C
752. O
753. E
754. G
755. M
756. J
757. H
758. F
759. A
760. D

HISTOLOGY

761. D
762. B
763. A
764. D
765. A
766. C
767. B
768. A
769. B
770. D
771. C
772. C
773. B
774. E
775. A
776. C
777. A
778. B
779. B
780. A
781. C
782. C
783. E
784. B
785. C
786. A, E
787. E
788. D
789. E
790. A, B, C
791. A, B, C, D, E
792. B, C
793. A, B
794. B
795. B, C, D
796. E
797. A, C, E
798. A, B. E
799. C, E
800. A, E
801. A, C
802. E
803. A, C, D, E
804. E
805. E
806. A
807. E
808. A, C, D, E
809. E
810. A, C, E
811. E
812. A
813. C
814. C
815. A
816. D
817. D
818. B
819. A
820. D
821. D
822. E
823. D
824. F
825. A
826. E
827. F
828. B
829. C
830. E
831. D
832. A
833. D
834. E
835. D
836. C
837. A
838. A
839. B
840. C
841. D
842. B
843. D
844. A
845. B
846. C
847. C
848. B
849. A
850. A
851. C
852. A
853. A
854. A
855. C
856. C
857. B
858. C
859. D
860. C
861. A
862. C
863. E
864. D
865. E
866. A

HISTOLOGY

867. B
868. C
869. D
870. C
871. C
872. B
873. E
874. D
875. D
876. E
877. D
878. E
879. A
880. A
881. B
882. B
883. E
884. B
885. C
886. A
887. B
888. A
889. A
890. A
891. B
892. A
893. A
894. B
895. C
896. A
897. B
898. C
899. C
900. A
901. A
902. A
903. C
904. A
905. C
906. A
907. C
908. B
909. A
910. A
911. E-1
912. A-1
913. B-2
914. C-1
915. D-3
916. A-2
917. B-2
918. A-3
919. B-2
920. D-1
921. C-1
922. E-2
923. E-2
924. E-3
925. A-3
926. B-1
927. C-2
928. C-3
929. D-1
930. E-3
931. A-1
932. B-2
933. C-3
934. D-3
935. E-2
936. B, D, A, E, C
937. A, B, C, D, E
938. C, B, A, D, E
939. E, C, B, D, A
940. B, A, D, C, E
941. D, B, C, A, E
942. E, B, C, D, A
943. A, B, C, D, E or E, D, C, B, A
944. B, D, E, A, C
945. C, D, E, A, B
946. B
947. A
948. C
949. D
950. E
951. B
952. C
953. E
954. A
955. D
956. B
957. D
958. C
959. E
960. A
961. A
962. E
963. B
964. C
965. D
966. A
967. B
968. C
969. D
970. E
971. B
972. E
973. A
974. M
975. H
976. K
977. C
978. D
979. I
980. F
981. O
982. A
983. K
984. E
985. G
986. B
987. M
988. F
989. I
990. H
991. B
992. O
993. A
994. L
995. C
996. D
997. F
998. I
999. N
1000. G
1001. C
1002. A
1003. F
1004. E
1005. J
1006. M
1007. L
1008. K
1009. D
1010. B
1011. C
1012. H
1013. D
1014. J
1015. L
1010. M
1017. A
1018. I
1019. B
1020. G

NEUROANATOMY

1021. A
1022. C
1023. C
1024. C
1025. E
1026. B
1027. D
1028. B
1029. B
1030. E
1031. C
1032. A
1033. C
1034. E
1035. B
1036. D
1037. B
1038. A
1039. C
1040. E
1041. B
1042. A
1043. C
1044. A
1045. E
1046. B
1047. B
1048. E
1049. A
1050. C
1051. C
1052. D
1053. A
1054. D
1055. C
1056. B
1057. A
1058. C
1059. B
1060. C
1061. E
1062. D
1063. D
1064. A
1065. B
1066. C
1067. A
1068. E
1069. C
1070. A
1071. B
1072. D
1073. E
1074. D
1075. B
1076. A
1077. A
1078. B
1079. C
1080. B
1081. E
1082. D
1083. C
1084. B
1085. C
1086. O
1087. J
1088. A
1089. E
1090. K
1091. H
1092. I
1093. N
1094. L
1095. G
1096. A
1097. I
1098. D
1099. J
1100. F
1101. B
1102. K
1103. H
1104. C
1105. E
1106. F
1107. A
1108. H
1109. D
1110. J
1111. I
1112. B
1113. L
1114. G
1115. E
1116. A, B, C, D
1117. B, C
1118. C, D, E
1119. C, D
1120. C, D
1121. A, B, C
1122. A, B, C, D, E
1123. A, E
1124. B, C, D
1125. D
1126. A, D
1127. A, B, C
1128. B, E
1129. E
1130. A, B, C, D, E
1131. C, D, E
1132. B, C
1133. B, C
1134. A, B, C, D, E
1135. A, B, C
1136. B, C
1137. A
1138. E
1139. A
1140. D
1141. A, D
1142. A, B, C
1143. D
1144. C, D, E
1145. E
1146. E
1147. B, D
1148. A
1149. A, B
1150. A, B, C
1151. A, B, C, D, E
1152. E
1153. A, B
1154. D
1155. A, C, D
1156. C
1157. C
1158. A
1159. B
1160. C
1161. D
1162. B
1163. C
1164. A
1165. D
1166. B
1167. C
1168. C
1169. D
1170. A
1171. C
1172. A
1173. C
1174. D
1175. D
1176. B
1177. A
1178. D
1179. E

NEUROANATOMY

1180. C
1181. C
1182. A
1183. B
1184. D
1185. E
1186. E
1187. K
1188. C
1189. G
1190. A
1191. F
1192. M
1193. H
1194. O
1195. B
1196. A
1197. E
1198. C
1199. C
1200. B
1201. D
1202. C
1203. D
1204. A
1205. E
1206. B
1207. C
1208. A
1209. B
1210. E
1211. A
1212. E
1213. D
1214. C
1215. A
1216. B
1217. B
1218. E
1219. B
1220. C
1221. D
1222. E
1223. D
1224. C
1225. C
1226. B
1227. D
1228. A
1229. B
1230. C
1231. B
1232. E
1233. C
1234. A
1235. D
1236. A-3
1237. C-2
1238. D-1
1239. E-2
1240. B-3
1241. B-1
1242. C-2
1243. D-3
1244. A-1
1245. A-1
1246. C
1247. B
1248. D
1249. E
1250. A
1251. B
1252. E
1253. D
1254. C
1255. A
1256. E
1257. J
1258. A
1259. F
1260. O
1261. G
1262. B
1263. I
1264. K
1265. D
1266. A, C, E
1267. A, E
1268. A, C
1269. B, C, D, E
1270. C, D
1271. A, B, C, D
1272. B, C, D
1273. D
1274. A, B
1275. B, C
1276. A, D
1277. A, B, C, D
1278. A, B, C
1279. D
1280. A, B, C, D, E
1281. C, D, E
1282. A, C, D
1283. A, C
1284. A, B, C
1285. A, B, C, D, E
1286. D
1287. A
1288. C
1289. C, D
1290. A, C, D
1291. A
1292. A, B, C
1293. C, D, E
1294. B, C, D
1295. B
1296. D
1297. B
1298. E
1299. A, D
1300. B
1301. A, B, C, D, E
1302. D
1303. B
1304. B, C
1305. D
1306. D
1307. A
1308. A
1309. F
1310. B
1311. E
1312. C
1313. E
1314. D
1315. D
1316. A
1317. F
1318. B
1319. C
1320. E
1321. D
1322. E
1323. A
1324. D
1325. F
1326. O
1327. A
1328. B
1329. G
1330. C
1331. F
1332. M
1333. I
1334. H
1335. E
1336. E, B, C, A, D
1337. B, C, A, D, E
1338. A, B, C, D, E

NEUROANATOMY

1339. C, B, D, A, E
1340. A, E, B, D, C
1341. E, D, C, B, A
1342. B, E, D, A, C
1343. C, A, B, E, D
1344. D, B, A, E, C
1345. B, E, C, A, D
1346. A, E, C, B, D
1347. C, A, D, B, E
1348. B, A, D, C, E
1349. D, B, A, E, C
1350. A, B, C, D, E
1351. C
1352. A
1353. E
1354. D
1355. E
1356. B
1357. A
1358. E
1359. B
1360. D
1361. B
1362. C
1363. A
1364. C
1365. D
1366. C
1367. A
1368. B
1369. B
1370. A
1371. C
1372. B
1373. D
1374. D
1375. E
1376. A
1377. B
1378. E
1379. E
1380. A
1381. E
1382. B
1383. D
1384. A
1385. B

EMBRYOLOGY AND TERATOLOGY

1386. C
1387. A
1388. C
1389. D
1390. A
1391. D
1392. C
1393. B
1394. C
1395. E
1396. A
1397. E
1398. B
1399. D
1400. C
1401. B
1402. A
1403. C
1404. B
1405. E
1406. A
1407. D
1408. B
1409. A
1410. B
1411. E
1412. C
1413. A
1414. B
1415. E
1416. B
1417. A
1418. C
1419. C
1420. E
1421. A
1422. E
1423. D
1424. D
1425. A
1426. E
1427. D
1428. B
1429. A
1430. A
1431. C
1432. C
1433. B
1434. D
1435. E
1436. D
1437. C
1438. D
1439. B
1440. E
1441. A
1442. C
1443. B
1444. C
1445. C
1446. D
1447. A
1448. C
1449. B
1450. A
1451. D
1452. B
1453. C
1454. B
1455. D
1456. A
1457. B
1458. D
1459. C
1460. E
1461. B
1462. C
1463. D
1464. E
1465. A
1466. D
1467. C
1468. E
1469. A
1470. B
1471. A
1472. B
1473. C
1474. D
1475. E
1476. C
1477. E
1478. A
1479. B
1480. D
1481. E
1482. D
1483. A
1484. M
1485. O
1486. N
1487. L
1488. C
1489. F
1490. G
1491. H
1492. C
1493. A
1494. F
1495. M
1496. O
1497. D
1498. D
1499. L
1500. J

OTHER BOOKS AVAILABLE

ITEMS	Code	Unit Price
MEDICAL EXAM REVIEW BOOKS		
Vol. 1 Comprehensive	101	$12.00
Vol. 2 Clinical Medicine	102	7.50
Vol. 2A Txtbk. Study Guide of Int. Med.	123	7.50
Vol. 2B Txtbk. Study Guide of Int. Med.	130	7.50
Vol. 3 Basic Sciences	103	7.50
Vol. 4 Obstetrics-Gynecology	104	7.50
Vol. 4A Textbk. Study Guide of Gynecology	152	7.50
Vol. 5 Surgery	105	7.50
Vol. 5A Textbk. Study Guide of Surgery	150	7.50
Vol. 6 Public Health & Prev. Medicine	106	7.50
Vol. 8 Psychiatry & Neurology	108	7.50
Vol.11 Pediatrics	111	7.50
Vol.12 Anesthesiology	112	7.50
Vol.13 Orthopaedics	113	10.00
Vol.14 Urology	114	10.00
Vol.15 Ophthalmology	115	10.00
Vol.16 Otolaryngology	116	10.00
Vol.17 Radiology	117	10.00
Vol.18 Thoracic Surgery	118	10.00
Vol.19 Neurological Surgery	119	15.00
Vol.20 Physical Medicine	128	10.00
Vol.21 Dermatology	127	10.00
Vol.22 Gastroenterology	141	10.00
Vol.23 Child Psychiatry	126	10.00
Vol.24 Pulmonary Diseases	143	10.00
Vol.25 Nuclear Medicine	133	10.00
Vol.26 Allergy	132	10.00
Vol.27 Plastic Surgery	129	10.00
Vol.28 Cardiovascular Diseases	138	10.00
Vol.29 Oncology	146	10.00
ECFMG Exam Review - Part One	120	7.50
ECFMG Exam Review - Part Two	121	7.50
BASIC SCIENCE REVIEW BOOKS		
Anatomy Review	201	7.00
Biochemistry Review	202	7.00
Digestive System Basic Sciences	215	7.00
Heart & Vascular Systems Basic Sciences	212	7.00
Microbiology Review	203	7.00
Nervous System Basic Sciences	210	7.00
Pathology Review	204	7.00
Pharmacology Review	205	7.00
Physiology Review	206	7.00
Respiratory System Basic Sciences	213	7.00
Urinary System B.Sci.	214	7.00
Anatomy Textbook Study Guide	124	7.00
Histology Textbook Study Guide	151	7.00
Medical Physiology Textbk. Study Guide	155	7.00
SPECIALTY BOARD REVIEW BOOKS		
Dermatology Specialty Board Review	311	10.00
Family Practice Specialty Board Review	309	10.00
Internal Medicine Specialty Board Review	303	10.00
Neurology Specialty Board Review	306	10.00
Obstetrics-Gynecology Spec. Bd. Review	304	10.00
Pathology Specialty Board Review	305	10.00
Pediatrics Specialty Board Review	301	10.00
Psychiatry Specialty Board Review	312	10.00
Surgery Specialty Board Review	302	10.00
The Otolaryngology Boards	313	10.00

ITEMS	Code	Unit Price
The Psychiatry Boards	307	$8.00
STATE BOARD REVIEW BOOKS		
Med. State Brd. Rev. - Basic Sciences	411	9.00
Med. State Brd. Rev. - Clinical Sciences	412	9.00
Cardiopulmonary Techn. Exam. Rev. - Vol. 1	473	7.50
Dental Exam. Review Book - Vol. 1	431	7.50
Dental Exam. Review Book - Vol. 2	432	7.50
Dental Exam. Review Book - Vol. 3	433	7.50
Dental Hygiene Exam. Review - Vol. 1	461	7.50
Emergency Med. Techn. Exam. Rev. - Vol. 1	465	7.50
Emergency Med. Techn. Exam. Rev. - Vol. 2	466	7.50
Immunology Exam. Review Book - Vol. 1	424	7.50
Inhalation Therapy Exam. Review - Vol. 1	471	7.50
Inhalation Therapy Exam. Review - Vol. 2	344	7.50
Laboratory Asst. Exam. Rev. Bk. - Vol. 1	455	7.50
Medical Librarian Exam. Rev. Bk. - Vol. 1	495	7.50
Medical Record Library Science - Vol. 1	496	7.50
Medical Techn. Exam. Review - Vol. 1	451	7.50
Medical Techn. Exam. Review - Vol. 2	452	7.50
Occupational Therapy Exam. Rev. - Vol. 1	475	7.50
Optometry Exam. Review	469	10.00
Pharmacy Exam. Review Book - Vol. 1	421	7.50
Physical Therapy Exam. Review - Vol. 1	481	7.50
Physical Therapy Exam. Review - Vol. 2	482	7.50
X-Ray Technology Exam. Rev. - Vol. 1	441	7.50
X-Ray Technology Exam. Rev. - Vol. 2	442	7.50
X-Ray Technology Exam. Rev. - Vol. 3	443	7.50
NURSING EXAM REVIEW BOOKS		
Vol. 1 Medical-Surgical Nursing	501	4.50
Vol. 2 Psychiatric-Mental Health Nursing	502	4.50
Vol. 3 Maternal-Child Health Nursing	503	4.50
Vol. 4 Basic Sciences	504	4.50
Vol. 5 Anatomy and Physiology	505	4.50
Vol. 6 Pharmacology	506	4.50
Vol. 7 Microbiology	507	4.50
Vol. 8 Nutrition & Diet Therapy	508	4.50
Vol. 9 Community Health	509	4.50
Vol.10 History & Law of Nursing	510	4.50
Vol.11 Fundamentals of Nursing	511	4.50
Practical Nursing Examination Rev. - Vol. 1	711	4.50
CASE STUDY BOOKS		
Allergy Case Studies	027	10.00
Cardiology Case Studies	001	10.00
Chest Diseases Case Studies	012	10.00
Child Psychiatry Case Studies	029	10.00
Cutaneous Medicine Case Studies	014	7.50
ECG Case Studies	003	7.50
Endocrinology Case Studies	008	10.00
Gastroenterology Case Studies	004	10.00
Hematology Case Studies	020	10.00
Infectious Diseases Case Studies	011	7.50
Neurology Case Studies	006	10.00
Otolaryngology Case Studies	021	10.00
Pediatric Hematology Case Studies	018	10.00
Pediatric Oculo-Neural Dis. Case Studies	023	10.00
Respiratory Care Case Studies	019	7.50
Urology Case Studies	017	10.00
MEDICAL OUTLINE SERIES		
Cancer Chemotherapy	631	10.00
Child Psychiatry	613	10.00

OTHER BOOKS AVAILABLE

ITEMS	Code	Unit Price
MEDICAL OUTLINE SERIES *(Cont'd.)*		
Endocrinology	614	$10.00
Histology	662	8.00
Otolaryngology	661	8.00
Psychiatry	621	8.00
Urology	611	8.00
SELF-ASSESSMENT BOOKS		
Self-Assess. Cur. Knldge - Biochemistry	266	7.50
S.A.C.K. in Cardiovascular Diseases	275	10.00
S.A.C.K. in Family Practice	261	10.00
S.A.C.K. in Infectious Diseases	263	10.00
S.A.C.K. in Internal Medicine	257	10.00
S.A.C.K. in Neurology	254	10.00
S.A.C.K. for Nurse Anesthetist	715	7.50
S.A.C.K. in O.R. Techn.	474	7.50
S.A.C.K. in Otolaryngology	270	10.00
S.A.C.K. in Pathology	253	10.00
S.A.C.K. in Pediatrics	256	10.00
S.A.C.K. in Psychiatry	252	10.00
S.A.C.K. in Pulmonary Diseases	271	10.00
S.A.C.K. in Rheumatology	258	10.00
S.A.C.K. in Surgery	250	10.00
S.A.C.K. in Surgery for Family Physicians	259	10.00
S.A.C.K. in Urology	251	10.00
MEDICAL HANDBOOKS		
Medical Emergencies	635	8.00
Neurology	604	8.00
Obstetrical Emergencies	634	8.00
Ophthalmologic Emergencies	633	8.00
Pediatric Anesthesia	637	8.00
Pediatric Neurology	636	10.00
PRACTICAL POINTS BOOKS		
In Anesthesiology	700	10.00
In Gastroenterology	733	7.00
In Pediatrics	702	10.00
PRACTITIONERS GUIDES		
OB-Gynecology Disorders	704	10.00
Ophthalmologic Disorders	703	10.00
JOURNAL ARTICLE COMPILATIONS		
Ambulance Service Journal Articles	517	10.00
Blood Banking & Immunohemat. Jour. Art.	798	10.00
Emergency Room Journal Articles	795	8.00
Hodgkin's Disease Journal Articles	515	12.00
Hosp. & Inst. Eng. & Maintenance J. Art.	793	8.00
Hosp. Electronic Data Process. J. Art.	791	8.00
Hosp. Pharmacy Journal Articles	799	10.00
Hosp. Security & Safety Journal Articles	796	8.00
Human Cytomegalovirus Journal Articles	522	15.00
Immunosuppressive Therapy Journal Art.	526	20.00
Institutional Laundry Journal Articles	789	8.00
Lithium & Psychiatry Journal Articles	520	15.00
Psychosomatic Medicine Current J. Art.	788	12.00

ITEMS	Code	Unit Price
Outpatient Services Journal Articles, 2nd Ed.	797	$10.00
Outpatient Services Journal Articles, *1st Ed.*	794	8.00
Selected Papers in Inhalation Therapy	523	10.00
TYPIST HANDBOOKS		
Medical Typist's Guide for Hx & Phys.	976	4.50
Radiology Typist Handbook	981	4.50
Surgical Typist Handbook	991	4.50
Transcribers Guide to Med. Terminology	973	4.50
ESSAY Q. & A. REVIEW BOOKS		
Blood Banking Principles Rev.	339	8.00
Cardiology Review	337	10.00
Colon & Rectal Surg. Cont. Ed. Rev.	338	10.00
Neurology Review	345	10.00
Obstet. Nursing Cont. Ed. Rev.	350	5.00
Ophthalmology Review	347	10.00
Orthopedics Review	349	10.00
Psychiatry Cont. Ed. Rev.	352	10.00
Psych./Mental Hlth. Nursing Cont. Ed. Rev.	351	5.00
OTHER BOOKS		
Acid Base Homeostasis	601	4.00
Allergy Annual Review	325	12.00
Bailey & Love's Short Practice of Surgery	900	20.00
Benign & Malignant Bladder Tumors	932	15.00
Blood Groups	860	2.50
Clinical Diagnostic Pearls	730	4.50
Concentrations of Solutions	602	3.00
Critical Care Manual	983	10.00
Cryogenics in Surgery	754	24.00
Diagnosis & Treatment of Breast Lesions	748	15.00
Emergency Care Manual	984	7.50
English-Spanish Guide for Med. Personnel	721	2.50
Fundamental Orthopedics	603	4.50
Guide to Medical Reports	962	4.50
Human Anatomical Terminology	982	3.00
Illustrated Laboratory Techniques	919	10.00
Introduction to Acupuncture	753	5.00
Introduction to Blood Banking	975	8.00
Introduction to the Clinical History	729	3.00
Math for Med Techs	964	7.00
Multilingual Guide for Medical Personnel	961	2.50
Neoplasms of the Gastrointestinal Tract	736	20.00
Neurophysiology Study Guide	600	7.00
Outpatient Hemorrhoidectomy Lig. Tech.	752	12.50
Profiles in Surgery, Gynec. & Obstetrics	963	5.00
Radiological Physics Exam. Review	486	10.00
Skin, Heredity & Malignant Neoplasms	744	20.00
Testicular Tumors	743	20.00
Tissue Adhesives in Surgery	756	24.00
Understanding Hematology	977	8.00